Global Solution
Curves for Semilinear
Elliptic Equations

Global Solution Curves for Semilinear Elliptic Equations

Philip Korman

University of Cincinnati, USA

NEW JERSEY · LONDON · SINGAPORE · BEIJING · SHANGHAI · HONG KONG · TAIPEI · CHENNAI

Published by

World Scientific Publishing Co. Pte. Ltd.

5 Toh Tuck Link, Singapore 596224

USA office: 27 Warren Street, Suite 401-402, Hackensack, NJ 07601

UK office: 57 Shelton Street, Covent Garden, London WC2H 9HE

British Library Cataloguing-in-Publication Data
A catalogue record for this book is available from the British Library.

ISBN-13 978-981-4374-34-7
ISBN-10 981-4374-34-2

Printed in Singapore.

Preface

The book reviews some recent results on global solution curves and *exact* multiplicity of positive solutions for the semilinear Dirichlet problem on a bounded domain D (here $u = u(x)$, $x \in D \subset R^n$)

$$\Delta u + \lambda f(u) = 0 \text{ in } D, \quad u = 0 \text{ on } \partial D, \tag{0.1}$$

depending on a positive parameter λ. Our goal is to gain a complete understanding of the solution curves $u = u(x, \lambda)$, and to determine how many *exactly* positive solutions this problem has at any particular value of λ. These goals are too ambitious at present for a general domain D, so most of the results are for balls in R^n, and the most detailed results are obtained for the case $n = 1$. By the classical theorem of B. Gidas, W.-M. Ni and L. Nirenberg [63], any positive solution of (0.1) on any ball, say the unit ball, is necessarily radially symmetric, i.e., $u = u(r)$, with $r = |x|$, and the problem (0.1) turns into an ODE

$$u''(r) + \frac{n-1}{r} u'(r) + \lambda f(u(r)) = 0 \text{ for } 0 < r < 1, \quad u'(0) = u(1) = 0. \tag{0.2}$$

After appearance of [63], the study of semilinear equations on balls began in earnest. (Of course, there were a number of results on radial solutions before, most notably another classic D.D. Joseph and T.S. Lundgren [75], but now radial solutions represented *all* positive solutions of the original PDE.) Soon, it was realized that the question of uniqueness of solutions is hard, even for relatively simple nonlinearities $f(u)$, like $f(u) = -u + u^p$, $f(u) = u + u^p$, and even the case $f(u) = u^p$ was non-trivial (uniqueness for the last case was already proved in [63]). After a great amount of effort by a number of people, uniqueness question for these equations was settled by the early nineties. We mention the following contributions: W.-M. Ni [137], W.-M. Ni and R.D. Nussbaum [138], M.K. Kwong [120], M.K. Kwong and

L. Zhang [122], L. Zhang [195], P. N. Srikanth [178], and Adimurthi and S. L. Yadava [3]. In case of multiple solutions, their *exact* multiplicity is also a hard question. When one applies variational or topological methods, a typical conclusion is that the problem has *at least* a certain number of solutions. By contrast, we wish to know the exact number of solutions, and how these solutions tie in together, i.e., what is the shape of global solution curves. With this information at hand, we can effectively compute all solutions, by using continuation methods (which are easy to implement). In fact, numerical computations had provided a major motivation for the bifurcation theory approach, presented in this book. This approach has been developed in the last twenty years by Y. Li, T. Ouyang, J. Shi and the present author.

What is the role of the parameter λ? Surely, it could be absorbed into the nonlinearity f. However, it is often helpful to have something "extra" in the statement of the problem. Consider, for example, the problem

$$\Delta u + 8u(u-1)(5-u) = 0 \ \text{ in } B, \ \ u = 0 \ \text{ on } \partial B, \tag{0.3}$$

where B is the unit ball in R^3. This problem has exactly two positive solutions, a fact which seems impossible to prove directly. We introduce a parameter λ, and consider

$$\Delta u + \lambda u(u-1)(5-u) = 0 \ \text{ in } B, \ \ u = 0 \ \text{ on } \partial B. \tag{0.4}$$

We now study *curves of solutions*, $u = u(x, \lambda)$. The advantage of this approach is that some parts of the solution curve are easy to understand, and it also becomes clear what are the tougher parts of the solution curve that we need to study - the *turning points*. For this example, it is easier to understand the larger stable solutions, which we then continue, by applying the Implicit Function Theorem (in Banach spaces), until a singular solution is reached, i.e., the Implicit Function Theorem is no longer applicable. We show that at singular solutions the Crandall-Rabinowitz Theorem (which we recall below) applies. It implies that either the solution curve continues in λ through the singular solution, or it just bends back (i.e., no secondary bifurcations or other eccentric behavior is possible). We then show that the unique global solution curve makes exactly one turn to the right, and the value of $\lambda = 8$ from (0.3) is to the right of the turn, which implies that the problem (0.3) has exactly two positive solutions.

The strategy of proving exact multiplicity results requires determining the direction of the turn at *singular* solutions (λ, u), i.e., solutions of (0.1) at which the corresponding linearized problem

$$\Delta w + \lambda f'(u)w = 0 \text{ in } D, \quad w = 0 \text{ on } \partial D \qquad (0.5)$$

admits a non-trivial solution. That, in turn, requires us to give conditions, ensuring that any non-trivial solution of (0.5) is of one sign. The most detailed results are obtained when one considers positive solutions of autonomous problems, i.e., when $f = f(u)$, but we also present both uniqueness and exact multiplicity results in case $f = f(r, u)$. (We obtain particularly detailed results for the one-dimensional case.) We distinguish between the case $f(0) \geq 0$, when positive solutions remain positive for all λ, and the case $f(0) < 0$, when solutions on a positive branch may become sign-changing, and symmetry breaking occurs. In addition to summarizing the work of the author and Y. Li, T. Ouyang, J. Shi over the last two decades, the book has a number of new results, or improved proofs of previously published results.

The paper of T. Ouyang and J. Shi [145] has a nice survey of earlier results, obtained using the bifurcation theory approach. We tried to avoid repetitions, concentrating either on different classes of problems, or problems where improvements of the results covered in [145] became available. The following are some of the improvements. Using the method of monotone separation of graphs, developed by L.A. Peletier and J. Serrin [148], [149], and M. Tang [181], we had obtained the exact shape of the solution curves for the nonlinearities including $f(u) = u^p + u^q$, $f(u) = (u + \gamma)^p$, and $f(u) = u^p + \gamma$ ($\gamma > 0$, a constant), for the natural range of powers $1 < p, q < \frac{n+2}{n-2}$. In [145], such results required that $1 < p, q < \frac{n}{n-2}$. By a different method, we have covered the concave-convex case, which includes $f(u) = u^p + u^q$, for $0 < p < 1 < q < \frac{n+2}{n-2}$. In case $f(0) < 0$, we study curves of positive solutions and symmetry breaking in a way which is simpler than in [145], and more consistent with the bifurcation approach. Among the topics of the present book, which were not covered in [145], we mention non-autonomous problems, equations with infinitely many turns, Hamiltonian systems, and problems on thin annular domains.

The bifurcation theory approach works also in the case of p-Laplacian, although there are many technical difficulties to overcome, and the work in this direction is still in progress. (I often wonder if "p" in p-Laplacian stands for pain.) In [101], building on the recent work of A. Aftalion and F. Pacella [11] and of a number of other people, we proved existence and uniqueness of positive radial solutions for a class of equations generalizing the model case

$$\Delta_p u - a(r)u^{p-1} + b(r)u^q = 0 \text{ in } B, \quad u = 0 \text{ on } \partial B,$$

where B is the unit ball in R^n, and Δ_p denotes the p-Laplace operator. We include an exposition of this result.

In the last chapter we give more detailed results for the one-dimensional case. Again, we tried to complement our earlier survey [92]. Among newer results, we mention an improvement of P. Korman and J. Shi [117] approach to stability and instability of solutions, exact multiplicity results for quasilinear equations (including the prescribed mean curvature equation), systems of equations and higher order equations. We also give an exposition of formulas, allowing one to compute the location and the direction of bifurcation, based on P. Korman, Y. Li, and T. Ouyang [111], and of related computer assisted proofs.

A unique feature of this book (and of the bifurcation theory approach) is integration of theoretical study of exact multiplicity results with numerical analysis and computation. In addition to computing of bifurcation diagrams, we present several computer assisted proofs of uniqueness and exact multiplicity theorems, as well as numerical computations using power series.

Philip L. Korman

Contents

Chapter 1

Curves of Solutions on General Domains

1.1 Continuation of solutions

The basic tool for continuation of solutions is the Implicit Function Theorem in Banach spaces. We present it here in the formulation of M.G. Crandall and P.H. Rabinowitz [39], see also L. Nirenberg [139].

Theorem 1.1. *Let X, Λ and Z be Banach spaces, and $f(x, \lambda)$ a continuous mapping of an open set $U \subset X \times \Lambda \to Z$. Assume that f has a Frechet derivative with respect to x, $f_x(x, \lambda)$, which is continuous on U. Assume that*

$$f(x_0, \lambda_0) = 0 \quad \text{for some } (x_0, \lambda_0) \in U.$$

If $f_x(x_0, \lambda_0)$ is an isomorphism (i.e., one-to-one and onto) of X onto Z, then there is a ball $B_r(\lambda_0) = \{\lambda : ||\lambda - \lambda_0|| < r\}$ and a unique continuous map $x(\lambda) : B_r(\lambda_0) \to X$, such that

$$f(x(\lambda), \lambda) \equiv 0, \quad x(\lambda_0) = x_0.$$

If f is of class C^p, so is $x(\lambda)$, $p \geq 1$.

In the conditions of the above theorem, we refer to (x_0, λ_0) as a *regular solution*, otherwise we call a solution to be *singular*.

We now consider positive solutions $u = u(x)$ of the semilinear elliptic boundary value problem

$$\Delta u + \lambda f(u) = 0 \quad \text{for } x \in D, \quad u = 0 \text{ on } \partial D, \tag{1.1}$$

depending on a positive parameter λ. We assume that D is a bounded region in R^n, whose boundary is of class C^2, and $f \in C^2(\bar{R}_+)$. Suppose that for $\lambda = \lambda_0$ the problem (1.1) has a solution $u_0(x)$. I.e., we have a

1

"point" (λ_0, u_0) on the solution curve. To study the solution curve through the point (λ_0, u_0), we need to consider the linearized problem for (1.1)

$$\Delta w(x) + \lambda_0 f'(u_0(x))w(x) = 0 \quad \text{for } x \in D, \quad w = 0 \text{ on } \partial D. \qquad (1.2)$$

We shall consider only the classical solutions of both problems (1.1) and (1.2), $u, w \in C^2(D) \cap C(\bar{D})$. If the problem (1.2) admits a non-trivial solution, we refer to $u_0(x)$ as a *singular solution* of (1.1), and (λ_0, u_0) as a *singular point*. If the point (λ_0, u_0) is *non-singular*, i.e., the problem (1.2) admits only the trivial solution $w(x) \equiv 0$, then the Implicit Function Theorem applies, and we can continue the solutions for both increasing and decreasing λ, obtaining locally a solution curve through the point (λ_0, u_0), see e.g., L. Nirenberg [139]. (Since the problem (1.2) admits only the trivial solution, the operator defined by its left hand side is one-to-one, and by the elliptic theory it is onto.) At a singular point (λ_0, u_0) practically anything imaginable may happen (several curves might pass through this point, or the picture may be even more complicated). The following theorem gives conditions for the solution set to be as simple as possible near a singular point (λ_0, u_0). It follows immediately from the abstract theorem of M.G. Crandall and P.H. Rabinowitz [40] (see Theorem 1.7 below), however we think it is instructive to give a self-contained proof.

Theorem 1.2. *Let (λ_0, u_0) be a singular point of (1.1). Assume that the solution space of the linearized problem (1.2) is one-dimensional, spanned by $w(x)$. Assume that the following non-degeneracy condition holds:*

$$\int_D f(u_0(x))w(x)\, dx \neq 0. \qquad (1.3)$$

Denote $Z = w^{\perp}$, the orthogonal (in $L^2(D)$) complement of $w(x)$ in $C^{2,\alpha}(D)$. Then the solutions of (1.1) near (λ_0, u_0) form a curve $(\lambda(s), u(s)) = (\lambda_0 + \tau(s), u_0 + sw + z(s))$, where $s \to (\tau(s), z(s)) \in \mathbf{R} \times Z$ is a continuously differentiable function near $s = 0$, and $\tau(0) = \tau'(0) = 0$, $z(0) = z'(0) = 0$.

The theorem implies that the solution set is a simple curve through (λ_0, u_0). This curve might continue to both sides of λ_0, or to one side of λ_0, in which case a turn occurs at λ_0. For the parameter s, one often uses the supremum norm of the solution $u(x)$.

Proof: We look for solution of (1.1) in the form $u = u_0 + sw + z(s)$, with $z(s) \in Z$. Since $\Delta u = \Delta u_0 + s\Delta w + \Delta z = -\lambda_0 f(u_0) - sf'(u)w + \Delta z$, the equation (1.1) becomes

$$F(z, \lambda, s) \equiv \Delta z - \lambda_0 f(u_0) - sf'(u_0 + sw + z)w \qquad (1.4)$$
$$+\lambda f(u_0 + sw + z) = 0.$$

Here $F(z, \lambda, s) : Z \times R^2 \to C^{0,\alpha}(D)$. We shall apply the Implicit Function Theorem to solve this equation for the pair (z, λ), as a function of s. The point $(z = 0, \lambda = \lambda_0, s = 0)$ satisfies the equation (1.4). We need to show that the Frechet derivative

$$F_{(z,\lambda)}(0, \lambda_0, 0)(\phi, \tau) = \Delta\phi + \lambda_0 f'(u_0)\phi + \tau f(u_0)$$

is one-to-one and onto.

If the Frechet derivative is zero, then $\phi \in Z$ satisfies

$$\Delta\phi + \lambda_0 f'(u_0)\phi = -\tau f(u_0) \text{ in } D, \quad \phi = 0 \text{ on } \partial D.$$

By the Fredholm alternative, the quantity on the right must be orthogonal to w, which implies that $\tau = 0$, in view of our condition (1.3). Then $\phi = cw$, since the solution space of the linearized problem (1.2) is one-dimensional, spanned by $w(x)$. But $\phi \in w^\perp$, which implies that $c = 0$.

Turning to the "onto" part, we need to show that for any $g(x) \in C^{0,\alpha}$ we can find a pair $(\phi \in Z, \tau \in R)$ solving

$$\Delta\phi + \lambda_0 f'(u_0)\phi + \tau f(u_0) = g(x) \text{ in } D, \quad \phi = 0 \text{ on } \partial D.$$

By the condition (1.3), we can choose τ, so that $g(x) - \tau f(u_0) \perp w(x)$. Then the above problem is solvable, and in fact it has a family of solutions $\phi = \phi_0 + cw$, with arbitrary constant c. We now choose c so that $\phi \perp w$, i.e., $\phi \in Z$. $\diamondsuit$

By the above theorem, $\tau(0) = \tau'(0) = 0$. If $\tau''(0) > 0$ (< 0) then $\tau(s) > 0$ (< 0), and hence $\lambda > \lambda_0$ $(\lambda < \lambda_0)$ near $s = 0$, and so the solution curve turns to the right (left) at (λ_0, u_0). The following formula is well known.

Theorem 1.3. *In the conditions of the Theorem 1.2, with $w(x)$ denoting a solution of (1.2), we have*

$$\tau''(0) = -\lambda_0 \frac{\int_D f''(u_0)w^3 \, dx}{\int_D f(u_0)w \, dx}. \qquad (1.5)$$

Proof: Differentiate the equation (1.1) twice in s,

$$\Delta u_{ss} + \lambda f'(u)u_{ss} + 2\tau' f'(u)u_s + \lambda f''(u)u_s^2 + \tau'' f(u) = 0 \quad \text{for } x \in D$$

$$u_{ss} = 0 \quad \text{on } \partial D.$$

Set here $s = 0$. By the Theorem 1.2, $\tau'(0) = 0$, and $u_s|_{s=0} = w$, and we obtain

$$\Delta u_{ss} + \lambda_0 f'(u_0)u_{ss} + \lambda_0 f''(u_0)w^2 + \tau''(0)f(u_0) = 0 \quad \text{for } x \in D$$

$$u_{ss} = 0 \quad \text{on } \partial D.$$

Multiplying this equation by w, the equation (1.2) by u_{ss}, subtracting and integrating, we obtain the formula (1.5).   ◇

The following result on bifurcation from the trivial solution will be sufficient for our purposes. More general situations are covered in M.G. Crandall and P.H. Rabinowitz [39]. We denote by $(\lambda_1, \varphi_1(x))$ the principal eigenpair of $\Delta u + \lambda u = 0$ in D, $u = 0$ on ∂D.

Theorem 1.4. *Let $f(u) \in C^2(\bar{R}_+)$, $f(0) = 0$, $f'(0) \neq 0$. Then positive solutions of (1.1) emerge near the point $(\frac{\lambda_1}{f'(0)}, 0)$. Moreover, all positive solutions of (1.1) near that point lie on a curve $(\lambda(s), u(s)) = (\frac{\lambda_1}{f'(0)} + \tau(s), s\varphi_1 + z(s))$, where $s \to (\tau(s), z(s))$ is a continuously differentiable function near $s = 0$, and $\tau(0) = 0$, $z(0) = z'(0) = 0$.*

The proof is similar to that of Theorem 1.2. The direction of bifurcation from zero is usually easy to compute. If $\frac{f(u)}{u} > f'(0)$ ($\frac{f(u)}{u} < f'(0)$) for small u, the bifurcation is to the left (right). To see that, just multiply the equation (1.1) by φ_1, and integrate over D.

We now discuss the situations when the non-degeneracy condition (1.3) holds. Assume first that D is a unit ball in R^n. According to the classical theorem of B. Gidas, W.-M. Ni and L. Nirenberg [63], any positive solution of (1.1) is radially symmetric, i.e., $u = u(r)$ (where $r = |x|$) satisfies

$$u''(r) + \frac{n-1}{r}u'(r) + \lambda f(u(r)) = 0 \quad \text{for } 0 \leq r < 1, \tag{1.6}$$

$$u'(0) = u(1) = 0.$$

Moreover, $u'(r) < 0$ for all $r \in (0,1)$. In view of C.S. Lin and W.-M. Ni [127], solutions of the corresponding linearized problem (1.2) are also radially symmetric, i.e., they satisfy

$$w''(r) + \frac{n-1}{r}w'(r) + \lambda f'(u(r))w(r) = 0 \quad \text{for } 0 \leq r < 1, \tag{1.7}$$

$$w'(0) = w(1) = 0.$$

Sign-changing solutions of (1.1) are not necessarily radially symmetric, however we may still consider radially symmetric sign-changing solutions, an important special class of solutions. The following theorem was first proved in P. Korman [79].

Theorem 1.5. *Let $u(r)$ be a singular solution of (1.6) (positive or sign-changing), i.e., the problem (1.7) admits a non-trivial solution. Then*

$$\int_0^1 f(u(r))w(r)\,r^{n-1}\,dr = \frac{1}{2\lambda}u'(1)w'(1)\,. \tag{1.8}$$

Proof: We consider the function

$$\zeta(r) = r^n\left[u'w' + \lambda f(u)w\right] + (n-2)r^{n-1}u'w,$$

introduced by M. Tang [182]. Using the equations (1.6) and (1.7), compute

$$\zeta'(r) = 2\lambda f(u)wr^{n-1}\,. \tag{1.9}$$

Integrating over the interval $(0,1)$, we conclude (1.8). $\diamondsuit$

We have $w'(1) \neq 0$ for any non-trivial solution of (1.7), in view of uniqueness for initial value problems. If we assume that

$$u'(1) \neq 0, \tag{1.10}$$

then $\int_0^1 f(u(r))w(r)\,r^{n-1}\,dr \neq 0$, which is the non-degeneracy condition (1.3). If $f(0) \geq 0$, then any positive solution of (1.6) satisfies $u'(1) < 0$, by the Hopf's boundary lemma (see e.g., M. Renardy and R.C. Rogers [157]), and the non-degeneracy condition (1.3) holds.

For general domains one has the following result. Here $n(x)$ denotes the unit normal vector at $x \in \partial D$, pointing outside.

Theorem 1.6. *Suppose that D is a bounded domain with C^2 boundary, such that $x \cdot n(x) > 0$ for all $x \in \partial D$. Assume that at some solution $(\lambda, u(x))$ of (1.1) the linearized problem (1.2) admits a positive solution $w(x)$. Then the non-degeneracy condition (1.3) holds.*

With an extra condition, $f(0) \geq 0$, this theorem was proved independently by T. Ouyang and J. Shi [144], and K.J. Brown [25], and a simpler proof was given subsequently by P. Korman [86]. Later, L. Nirenberg [140] has observed that the condition $f(0) \geq 0$ is not necessary, and the details were written up in P. Korman [90]. The domain D described in the Theorem 1.6 is called *star-shaped*.

Observe that if u is a singular solution of (1.1), then zero is an eigenvalue of the corresponding linearized operator, i.e., $\mu = 0$ is an eigenvalue of

$$\Delta w(x) + \lambda f'(u(x))w + \mu w = 0 \quad \text{for } x \in D, \quad w = 0 \text{ on } \partial D, \qquad (1.11)$$

and the corresponding eigenfunction w is then a solution of (1.2). The assumption that $w > 0$ in the Theorem 1.6 implies that $\mu = 0$ is the first (principal) eigenvalue of (1.11).

Proof of Theorem 1.6. We assume that D is *not* a ball, since on a ball even more general result is true (see above). We claim that

$$\int_D f(u)w \, dx = -\frac{1}{2\lambda} \int_{\partial D} \frac{\partial w}{\partial n} |\nabla u| \, (x \cdot n(x)) \, ds. \qquad (1.12)$$

Indeed, the function $v = x \cdot \nabla u$ satisfies

$$\Delta v(x) + \lambda f'(u(x))v = -2\lambda f(u(x)).$$

Combining this with the linearized problem (1.2),

$$\int_D f(u)w \, dx = \frac{1}{2\lambda} \int_{\partial D} v \frac{\partial w}{\partial n} \, ds. \qquad (1.13)$$

Since ∂D is a level set of $u(x)$, it follows that $\nabla u = -|\nabla u|n$, and hence $v = -|\nabla u| \, (x \cdot n)$. Using this in (1.13), we conclude (1.12).

Clearly on ∂D, $\frac{\partial u}{\partial n} \leq 0$, while by the Hopf's lemma, $\frac{\partial w}{\partial n} < 0$. If $\frac{\partial u}{\partial n}$ is negative somewhere on ∂D, then $|\nabla u| > 0$ there, and hence the integral on the right hand side of (1.12) is positive, and we are done. So suppose $\frac{\partial u}{\partial n} \equiv 0$ on ∂D. I.e., $u(x)$ satisfies the equation (1.1), with over-determined boundary conditions $u = \frac{\partial u}{\partial n} = 0$ on ∂D. But then by the classical results of J. Serrin [166], and of B. Gidas, W.-M. Ni and L. Nirenberg [63], D is a ball, contrary to our assumption (see also Theorem 2.18 in J. Bebernes and D. Eberly [17]). $\diamondsuit$

Our requirement that any non-trivial solution of the linearized problem (1.2) is positive, represents a very strong restriction. But it holds in the following case (pointed out previously in K.J. Brown [25]). Let $\lambda_1 < \lambda_2 \leq \lambda_3 \leq \cdots$ denote the eigenvalues of $\Delta u + \lambda u = 0$ in D, $u = 0$ on ∂D. Observe that any non-trivial solution of the linearized problem (1.2) is an eigenfunction of (1.11), corresponding to the eigenvalue $\mu = 0$. If one assumes that

$$\lambda f'(u(x)) < \lambda_2, \quad \text{for all } x \in D, \qquad (1.14)$$

then zero is the first eigenvalue, i.e., $w(x) > 0$ (since the eigenvalues of $\Delta u + a(x)u$ are monotone in $a(x)$). Condition (1.14) is also very restrictive, in fact it will almost never hold for an entire solution curve. However, it may be possible to verify this condition for a part of a solution curve, where $u(x)$ (and hence $f'(u(x))$) is bounded. Computer assisted uniqueness and exact multiplicity results can be produced this way, similarly to a recent paper of P.J. McKenna et al [132]. That paper proves uniqueness of positive solution for the problem

$$\Delta u + \lambda u + u^2 = 0 \quad \text{for } x \in D, \quad u = 0 \text{ on } \partial D, \qquad (1.15)$$

where D is a unit square $(0,1) \times (0,1)$ in R^2, and $\lambda \in [0, \lambda_1)$. Of course, the solution curve extends to negative λ also. Solutions become large as $\lambda \to -\infty$, and computers cannot help us. In case D is a ball, uniqueness holds for all $\lambda \in (-\infty, \lambda_1)$ (see Theorem 2.14 below).

Is it possible to get around the condition $w > 0$, i.e., the assumption that $\mu = 0$ is the first eigenvalue of (1.11)? As we saw above, this condition is not necessary for balls in R^n. In the next section we observe that situation is similar for symmetric domains in R^2, following M. Holzmann and H. Kielhofer [72], and L. Damascelli, M. Grossi and F. Pacella [44]. Then we show that an elegant argument of C.S. Lin [126] implies that for convex domains in R^2 the non-degeneracy condition holds if $\mu = 0$ is either first or second eigenvalue. What about domains that are not star-shaped? This appears to be a hard question. Let us consider positive and decreasing radial solutions on an annulus $A : a < r < b$, i.e.,

$$u''(r) + \frac{n-1}{r} u'(r) + \lambda f(u) = 0 \quad r \in (a,b), \quad u(a) = u(b) = 0.$$

Integrating (1.9), we have

$$\int_A f(u(x))w(x)\,dx = \frac{\omega_n}{2\lambda} \left(b^n u'(b) w'(b) - a^n u'(a) w'(a) \right).$$

If $w > 0$, it is not at all clear if this quantity is non-zero. On the other hand, the non-degeneracy condition (1.3) does hold if $\mu = 0$ is either second, fourth or any other even eigenvalue, since then $w'(b)$ and $w'(a)$ have opposite sign.

We shall also need the original "abstract" version of the Crandall-Rabinowitz Theorem, which we review next.

Theorem 1.7. ([40]) *Let X and Y be Banach spaces. Let $(\overline{\lambda}, \overline{x}) \in \mathbf{R} \times X$ and let F be a continuously differentiable mapping of an open neighborhood of $(\overline{\lambda}, \overline{x})$ into Y. Let the null-space $N(F_x(\overline{\lambda}, \overline{x})) = \text{span}\,\{x_0\}$ be one-dimensional and codim $R(F_x(\overline{\lambda}, \overline{x})) = 1$. Let $F_\lambda(\overline{\lambda}, \overline{x}) \notin R(F_x(\overline{\lambda}, \overline{x}))$. If Z*

is a complement of span $\{x_0\}$ *in* X, *then the solutions of* $F(\lambda, x) = F(\overline{\lambda}, \overline{x})$ *near* $(\overline{\lambda}, \overline{x})$ *form a curve* $(\lambda(s), x(s)) = (\overline{\lambda} + \tau(s), \overline{x} + sx_0 + z(s))$, *where* $s \rightarrow (\tau(s), z(s)) \in \mathbf{R} \times Z$ *is a continuously differentiable function near* $s = 0$ *and* $\tau(0) = \tau'(0) = 0$, $z(0) = z'(0) = 0$.

We shall need the following particular case of the a priori estimates of B. Gidas and J. Spruck [64], in order to show that solution curves do not go to infinity, when the parameter λ belongs to a compact subset of $(0, \infty)$.

Theorem 1.8. *Let* $u(x)$ *be any positive classical solution of the problem*

$$\Delta u + f(x, u) = 0 \ \text{in} \ D, \quad u = 0 \ \text{on} \ \partial D.$$

Assume that $\lim_{u \to \infty} \frac{f(x,u)}{u} = a(x)$ *exists, uniformly in* $x \in \bar{D}$, *for some* $1 < p < \frac{n+2}{n-2}$, *and the function* $a(x)$ *is continuous and strictly positive on* $\bar{D}$. *There there is a constant* K, *such that*

$$u(x) \le K \ \text{for all} \ x \in D.$$

We conclude this section with the following general observations. We denote $F(u) = \int_0^u f(t)\,dt$, as usual.

Theorem 1.9. *Let* $D \subset R^n$ *be a bounded convex domain,* $f(u) \in C^1(\bar{R}_+)$. *Let* $u(x) \in C^3(D) \cap C^2(\bar{D})$ *be a positive solution of (1.1), and let* $\alpha = \max_D u(x)$. *Then*

$$F(\alpha) > 0. \tag{1.16}$$

This will follow immediately from the following maximum principle of L. Payne (see Corollary 5.1 in R. Sperb [177], and Theorem 2.3 in Z. Ding et al [49]).

Theorem 1.10. *Let* $D \subset R^n$ *be a bounded convex domain,* $f(u) \in C^1(\bar{R}_+)$. *If* $u \in C^3(D) \cap C^2(\bar{D})$ *is a solution of (1.1) (not necessarily positive), then the function* $P(x) = \frac{1}{2}|\nabla u(x)|^2 + F(u(x))$ *assumes its maximum over* $\bar{D}$ *only inside* D, *at a point where* $\nabla u(x) = 0$.

Proof of the Theorem 1.9 On the boundary ∂D, the function $P(x) = \frac{1}{2}|\nabla u(x)|^2 \ge 0$. Hence at the point of maximum, $P(x) = F(\alpha) > 0$. $\diamondsuit$

It was observed in R. Sperb [177] that for any positive solution of (1.1)

$$\lambda < \lambda_1 \sup_{s>0} \frac{s}{f(s)}, \tag{1.17}$$

provided that $f(u) > 0$ for $u > 0$. (Just multiply the equation (1.1) by the principal eigenfunction of the Laplacian, and integrate.) In particular, for a superlinear at infinity $f(u)$, the solution curve cannot extend to infinity.

1.2 Symmetric domains in R^2

In this section we show that the non-degeneracy condition (1.3) holds for symmetric domains in R^2, without any restriction on w, similarly to balls in R^n. This result was implicit in M. Holzmann and H. Kielhofer [72], see also P. Korman [90].

Theorem 1.11. *Let D be a bounded domain in two dimensions, which is symmetric with respect to both x and y axes, and convex in both x and y directions, with ∂D of class C^3. Assume that $f(u) \in C^1(\bar{R}_+)$ satisfies $f(0) \geq 0$. Let $u(x,y)$ be a singular positive solution of (1.1). Then the non-degeneracy condition (1.3) holds, and hence the Theorem 1.2 applies.*

Prior to giving the proof, we recall some known facts, assuming that the conditions of the theorem hold. By the classical results of B. Gidas, W.-M. Ni and L. Nirenberg [63], any positive solution is an even function in both x and y, with $u_x < 0$ and $u_y < 0$ in the first quadrant $x, y > 0$, i.e., $u(0,0)$ gives the maximum value of the solution. Next we recall that the Theorem 2.1 of L. Damascelli, M. Grossi and F. Pacella [44] implies the following lemma.

Lemma 1.1. *In the conditions of the Theorem 1.11, any solution of the linearized problem (1.2) is also even in both x and y.*

This lemma can be seen as an analog of C.S. Lin and W.-M. Ni [127] result for balls. We shall also need the following fundamental lemma of M. Holzmann and H. Kielhofer [72].

Lemma 1.2. *([72]) In the conditions of the Theorem 1.11, there is no nontrivial solution $v(x,y)$ of the linearized equation*

$$\Delta v + \lambda f'(u)v = 0 \quad \text{for } x \in D, \tag{1.18}$$

that is even in both x and y, which satisfies $v(0,0) = 0$, and such that either $v = 0$ or $v < 0$ on ∂D.

Combining these lemmas, we see that any non-trivial solution $w(x,y)$ of the linearized problem (1.2), satisfies

$$w(0,0) \neq 0. \tag{1.19}$$

Proof of the Theorem 1.11 Observe that the solution set of the linearized problem (1.2) is one dimensional, since otherwise we could produce a non-trivial solution of (1.2) with $w(0,0) = 0$, in contradiction with (1.19). Assume that, on the contrary, $\int_D f(u)w\,dx dy = 0$. By the Fredholm alternative, we can find $v_1(x,y)$ which is a solution of

$$\Delta v_1 + \lambda f'(u(x,y))v_1 = -2\lambda f(u) \quad \text{for } x \in D, \quad v_1 = 0 \text{ on } \partial D. \quad (1.20)$$

We claim that $v_1(x,y)$ is even in both x and y. Indeed, if $v_1(x,y)$ was not even in x, then $v_1(-x,y)$ would be another solution of (1.20), and then $v_1(x,y) - v_1(-x,y)$ would be a solution of the linearized problem (1.2) that is not even in x, a contradiction with Lemma 1.1.

One checks that the function $v(x,y) = xu_x + yu_y$ also satisfies (1.20). Assume first that $v_1(0,0) \neq 0$. Consider $z(x,y) \equiv w + v - v_1$. By scaling w, we may achieve $z(0,0) = 0$. The function $z(x,y)$ is even in both x and y, and it satisfies the equation (1.18), and $z < 0$ on ∂D (by Hopf's lemma, since $f(0) \geq 0$). We have a contradiction with Lemma 1.2. In case $v_1(0,0) = 0$, we define $z = v - v_1$, and again we have $z(0,0) = 0$, while $z < 0$ on ∂D, in contradiction with Lemma 1.2. $\diamond$

We remark that the condition $f(0) \geq 0$ was one of the assumptions in the original paper [72]. We believe that this assumption can be avoided, similarly to the case of balls in R^n.

The analogy between symmetric domains in R^2 and balls in R^n is further underscored by the following result of M. Holzmann and H. Kielhofer [72].

Theorem 1.12. *In addition to the conditions of the Theorem 1.11, assume that $f \in C^2(\bar{R}_+)$ satisfies*

$$f(u) \geq 0 \ \text{for} \ u \in [0,b), \quad 0 < b \leq \infty.$$

Then the value of $p = u(0,0)$ can used as a global parameter, i.e., for any $p \in (0,b)$ there is at most one pair (λ, u) satisfying (1.1).

1.3   Turning points and the Morse index

For any solution $u(x)$ of (1.1), we may consider an eigenvalue problem

$$\Delta w(x) + \lambda f'(u(x))w + \mu w = 0 \quad \text{for } x \in D, \quad w = 0 \text{ on } \partial D. \quad (1.21)$$

According to the standard theory of elliptic operators, the eigenvalues of (1.21) form a sequence $\mu_1 < \mu_2 < \cdots < \mu_n < \cdots$, tending to infinity. The number of negative eigenvalues is called the *Morse index* of the solution $u(x)$. If $u(x)$ is a singular solution of (1.1), then $\mu = 0$ is an eigenvalue of

(1.21), and the corresponding eigenfunction w is a solution of the linearized problem (1.2). The following result was implicit in K. Nagasaki and T. Suzuki [136].

Theorem 1.13. *Let $u_0(x)$ be a positive singular solution of (1.1), corresponding to $\lambda = \lambda_0$, and assume that the conditions of the Crandall-Rabinowitz Theorem 1.2 hold. Let $w(x)$ again denote solution of the linearized problem (1.2), computed in the direction the solution curve travels through the turning point, i.e., $w = u_s|_{s=0}$, and assume that*

$$\int_D f''(u_0(x))w^3(x)\,dx \neq 0. \tag{1.22}$$

Then as the solution curve passes through the point (λ_0, u_0), the Morse index of solution changes by one. Namely, it increases (decreases) by one if $\int_D f''(u_0(x))w^3(x)\,dx > 0$ (< 0).

Proof: Assume that zero is the k-th eigenvalue of (1.21), and let $w(x)$ denote the corresponding eigenfunction. According to the Theorem 1.2, near the point (λ_0, u_0) solutions are parameterized by s. Let $(\mu_k(s), \phi(s))$ denote the k-th eigenpair of (1.21) along the curve, i.e.,

$$\Delta\phi(s) + \lambda(s)f'(u(x,s))\phi(s) + \mu_k(s)\phi(s) = 0 \text{ in } D, \quad \phi(s) = 0 \text{ on } \partial D. \tag{1.23}$$

Differentiate the equation (1.23) in s:

$$\Delta\phi_s + \lambda f'(u)\phi_s + \lambda f''(u)u_s\phi + \lambda' f'(u)\phi + \mu_k'\phi + \mu_k\phi_s = 0.$$

We now set $s = 0$. Since $\mu_k(0) = 0$, $u_s|_{s=0} = w$, $\phi(0) = w$, $\lambda'(0) = 0$, and also $\lambda(0) = \lambda_0$, $u|_{s=0} = u_0$, we get

$$\Delta\phi_s + \lambda_0 f'(u_0)\phi_s + \lambda_0 f''(u_0)w^2 + \mu_k'(0)w = 0, \quad \text{for } x \in D$$

$$\phi_s = 0 \text{ on } \partial D.$$

Combining this with the linearized problem (1.2) at the point (λ_0, u_0), we conclude

$$\mu_k'(0) = -\frac{\lambda_0 \int_D f''(u_0)w^3\,dx}{\int_D w^2\,dx}.$$

If $\int_D f''(u_0)w^3\,dx > 0$, the k-th eigenvalue becomes negative after passing through the turning point (λ_0, u_0), i.e., the Morse index of solution is increased by one. $\diamond$

If the condition (1.22) holds, we call the critical point (λ_0, u_0) to be *non-degenerate*. Non-degenerate critical points persist under a small perturbation of the equation (1.1), which happens if, for example, this equation depends on a secondary parameter, in addition to λ, see J. Shi [169] or P. Korman, Y. Li and T. Ouyang [110]. In addition, the sign of the integral in (1.22) governs the direction of bifurcation, as we shall see later.

1.4   Convex domains in R^2

We now consider two-dimensional convex domains, i.e., $D \subset R^2$ and $u = u(x, y)$. Recall that Courant's Nodal Domain Theorem [38] asserts that the nodal set of the k-th eigenvalue of (1.21) divides the domain D into at most k regions. It follows that in two-dimensional case the nodal line of the second eigenfunction divides the region D into two regions, since only the principal eigenfunction is of one sign. In P. Korman [90], we had adapted an argument of C.S. Lin [126] to prove the following theorem.

Theorem 1.14. *Let D be a bounded convex domain in two dimensions. Assume $u(x, y)$ is a singular positive solution of (1.1), so that $\mu = 0$ is either first or second eigenvalue of (1.21). Then the non-degeneracy condition (1.3) holds.*

Proof: We may assume that D is not a ball, since on a ball even more general result is true. We may also assume that $\mu = 0$ is the second eigenvalue, since the case of the first eigenvalue is covered in the Theorem 1.6. (Clearly, a convex domain is star-shaped with respect to any of its points.)

Assume, on the contrary, that

$$\int_D f(u)w \, dxdy = 0. \tag{1.24}$$

The function $v(x, y) = xu_x + yu_y$ satisfies

$$\Delta v + \lambda f'(u)v = -2f(u). \tag{1.25}$$

Combining this with the linearized problem (1.2), we conclude

$$\int_{\partial D} v \frac{\partial w}{\partial n} \, ds = 0. \tag{1.26}$$

By Courant's nodal line theorem the nodal line $\{x \in D \mid w(x) = 0\}$ divides D into two sub-domains. According to the Proposition 2.4 in M. Holzmann and H. Kielhofer [72], there are only the following two possibilities to consider.

Case (i) The nodal line encloses a region inside D (i.e., either the nodal line lies strictly inside D or it touches ∂D at just one point). We choose as an origin any point inside D. We have $v \leq 0$ on ∂D. If $v \equiv 0$ on ∂D, then $u_x = u_y = 0$ on ∂D, and as before D is a ball, a contradiction. Hence $v < 0$ on some part of ∂D. By Hopf's lemma, $\frac{\partial w}{\partial n}$ is of one sign on ∂D, and so we have a contradiction in (1.26).

Case (ii) The nodal line intersects the boundary ∂D at two points, p_1, $p_2 \in \partial D$. Suppose first that the tangent lines to ∂D at p_1 and p_2 are not parallel. We choose origin O to be at their point of intersection (i.e., $x = y = 0$ at O). We decompose ∂D into two parts $\partial D = S_1 \cup S_2$, where S_1 is the part of ∂D lying inside the triangle Op_1p_2, and S_2 is its complement. Observe that ∇u gives direction of the inside pointing normal on ∂D. The vector (x, y) points inside D on S_1, and to outside of D on S_2. It follows that on S_1, $v = (x, y) \cdot \nabla u > 0$, while $v < 0$ on S_2. By Hopf's lemma, $\frac{\partial w}{\partial n}$ has one sign on S_1, and the opposite sign on S_2, again we have a contradiction in (1.26).

Assume now that the tangent lines at p_1 and p_2 are parallel, and go along the x-axis. Differentiating the equation (1.1), we have

$$\Delta u_x + \lambda f'(u)u_x = 0.$$

Combining this with linearized problem (1.2), we conclude

$$\int_{\partial D} u_x \frac{\partial w}{\partial n}\, ds = 0. \tag{1.27}$$

This time let S_1 be the part of ∂D to the left of the line p_1p_2, and S_2 the part to the right of the line. On S_2, $u_x < 0$, while $u_x > 0$ on S_1. The integrand in (1.27) is of one sign, a contradiction. $\diamond$

We remark that this theorem applies if e.g.,

$$\lambda f'(u) < \lambda_3$$

holds, which will usually hold for a part of solution curve.

We consider next positive solutions of

$$\Delta u + u^p = 0 \quad \text{for } x \in D, \quad u = 0 \text{ on } \partial D, \tag{1.28}$$

where $p > 1$, and D is a convex domain in two dimensions. If $u(x)$ is a singular solution, then the linearized problem

$$\Delta w + pu^{p-1}w = 0 \quad \text{for } x \in D, \quad w = 0 \text{ on } \partial D, \tag{1.29}$$

has non-trivial solutions, i.e., $\mu = 0$ is an eigenvalue of (1.21). The following is the crucial lemma of C.S. Lin [126]. We slightly simplify the original proof (by considering the function z below).

Lemma 1.3. $\mu = 0$ *is not the second eigenvalue of (1.21).*

Proof: The function $z = xu_x + yu_y + \frac{2}{p-1}u$ satisfies

$$\Delta z + pu^{p-1}z = 0 \,.$$

Combining this equation with the linearized problem (1.29),

$$\int_{\partial D} z \frac{\partial w}{\partial n} \, ds = 0 \,. \tag{1.30}$$

On the boundary ∂D, z agrees with the function v from the relation (1.26), hence the argument used in the proof of Theorem 1.14 (due to C.S. Lin [126]) applies. $\diamond$

This lemma was used by C.S. Lin [126] to prove the following result (we rephrase the original statement).

Theorem 1.15. *The problem (1.28) has exactly one positive solution of Morse index one.*

This result of C.S. Lin [126] remains the only known uniqueness result for general convex domains. While it is known what can go wrong for non-convex domains (the problem (1.28) may have multiple solutions for dumb-bell domains, according to the well-known results of E.N. Dancer [48]), it is not clear how to exploit the fact that the domain is convex.

For symmetric domains in R^2, there is the following unconditional uniqueness result of L. Damascelli, M. Grossi and F. Pacella [44].

Theorem 1.16. *([44]) Let D be a bounded domain in two dimensions, which is symmetric with respect to both x and y axes, and convex in both x and y directions, with ∂D of class C^3. Then the problem (1.28) has exactly one positive solution.*

We shall give a different proof of this result, retaining however the following crucial ingredient from [44].

Lemma 1.4. *([44]) In the conditions of the Theorem 1.16, the nodal line $w(x, y) = 0$ of any non-trivial solution of (1.29) encloses a region inside D.*

Lemma 1.5. *In the conditions of the Theorem 1.16, $\mu = 0$ is not the second eigenvalue of (1.21).*

Proof: We use again $z = xu_x + yu_y + \frac{2}{p-1}u$, for which the relation (1.30) holds. We may assume that D is not a ball, since for balls the Theorem 1.16 is well known, see [63]. Choose as an origin any point inside D. We have $z \leq 0$ on ∂D. If $z \equiv 0$ on ∂D, then $u_x = u_y = 0$ on ∂D, and as before D is a ball, a contradiction. Hence $z < 0$ on some part of ∂D. By Lemma 1.4, w is of one sign near ∂D, and then by Hopf's lemma $\frac{\partial w}{\partial n}$ is of one sign on ∂D, and so we have a contradiction in (1.30). $\quad\diamond$

Proof of the Theorem 1.16 We embed (1.28) into a parameter dependent problem

$$\Delta u + \lambda u^p = 0 \quad \text{for } x \in D, \quad u = 0 \text{ on } \partial D, \qquad (1.31)$$

where λ is a positive parameter. For any fixed λ, this problem has a positive solution, produced by the Mountain Pass Lemma, which by [63] is symmetric in both x and y. The Morse index of this solution is one, see e.g., C.S. Lin [126]. If solution is non-singular, we can continue it in λ by the Implicit Function Theorem. Let us continue this curve for decreasing λ. If a singular solution is reached, $\mu = 0$ is either the first or the second eigenvalue of (1.21). It cannot be the second eigenvalue by Lemma 1.5. So assume that $\mu = 0$ is the first eigenvalue of (1.21), i.e., $w > 0$ in D. By Theorem 1.11, we have a curve of solutions passing through the critical point. By Theorem 1.3, this curve has to make a turn to the left, which is impossible, since the curve has arrived from the right. Hence, no turns may occur. Similarly, we show that no turns happen if we continue the solution curve for increasing λ (if a turn to the left occurred, the curve would have no place to go). By scaling, we see that $u \to 0$ as $\lambda \to \infty$, and $u \to \infty$ as $\lambda \to 0$. Since this curve exhausts all possible maximum values $u(0,0)$, this curve exhausts the solution set of the problem (1.31), by the Theorem 1.12. At $\lambda = 1$, we get the unique solution of (1.28). $\quad\diamond$

We see that the Lemma 1.4 is crucial for uniqueness for (1.28). It is natural to ask: what are other domains for which the Lemma 1.4 holds?

1.5   Pohozaev's identity and non-existence of solutions for elliptic systems

Any solution $u(x)$ of semilinear Dirichlet problem on a bounded domain $D \subset R^n$

$$\Delta u + f(u) = 0 \text{ in } D, \quad u = 0 \text{ on } \partial D \qquad (1.32)$$

satisfies the well-known Pohozaev's identity [151]

$$\int_D [2nF(u) + (2 - n)uf(u)] \, dx = \int_{\partial D} (x \cdot \nu)|\nabla u|^2 \, dS. \qquad (1.33)$$

Here $F(u) = \int_0^u f(t)\,dt$, and ν is the unit normal vector on ∂D, pointing outside. From the equation (1.32), $\int_D uf(u)\,dx = \int_D |\nabla u|^2\,dx$, which gives an alternative form of Pohozaev's identity:

$$\int_D \left[2nF(u) + (2-n)|\nabla u|^2\right] \, dx = \int_{\partial D} (x \cdot \nu)|\nabla u|^2 \, dS.$$

A standard use of this identity is to conclude that if D is a star-shaped domain with respect to the origin, i.e., $x \cdot \nu \geq 0$ for all $x \in \partial D$, and $f(u) = u|u|^{p-1}$, for some constant p, then the problem (1.32) has no non-trivial solution in the super-critical case, when $p > \frac{n+2}{n-2}$. We present next a proof of this identity, which appears a little more straightforward than the usual one, see e.g., L. Evans [54], and then use a similar idea for systems, to derive the well-known result of E. Mitidieri [135] and R.C.A.M. Van der Vorst [186] on non-existence of solutions for power nonlinearities, involving the critical hyperbola.

Let $z = x \cdot \nabla u = \Sigma_{i=1}^n x_i u_{x_i}$. It is easy to verify that z satisfies

$$\Delta z + f'(u)z = -2f(u). \qquad (1.34)$$

We multiply the equation (1.32) by z, and subtract from that the equation (1.34) multiplied by u, obtaining

$$\Sigma_{i=1}^n \left[(zu_{x_i} - uz_{x_i})_{x_i} + x_i \frac{\partial}{\partial x_i}(2F(u) - uf(u)) \right] = 2f(u)u. \qquad (1.35)$$

Clearly,

$$\Sigma_{i=1}^n x_i \frac{\partial}{\partial x_i}(2F - uf) = \Sigma_{i=1}^n \frac{\partial}{\partial x_i}[x_i(2F - uf)] - n(2F - uf).$$

We then rewrite (1.35)

$$\Sigma_{i=1}^n \left[(zu_{x_i} - uz_{x_i}) + x_i(2F(u) - uf(u)) \right]_{x_i} = 2nF(u) + (2 - n)uf(u). \qquad (1.36)$$

Integrating over D, we conclude Pohozaev's identity (1.33). (The only non-zero boundary term is $\Sigma_{i=1}^n \int_{\partial D} z u_{x_i} \nu_i \, dS$. Since ∂D is a level set of u, $\nu = \pm \frac{\nabla u}{|\nabla u|}$, i.e., $u_{x_i} = \pm |\nabla u| \nu_i$. Then $z = \pm (x \cdot \nu) |\nabla u|$, and $\Sigma_{i=1}^n u_{x_i} \nu_i = \pm |\nabla u|$.)

It appears natural to refer to (1.36) as a *differential form* of Pohozaev's identity. For radial solutions on a ball, the corresponding version of (1.36) played a crucial role in the study of exact multiplicity of solutions, see T. Ouyang and J. Shi [145], and also P. Korman [93]. These results will be presented in the next chapter.

Similar derivation shows that for non-autonomous problem

$$\Delta u + f(x, u) = 0 \text{ in } D, \quad u = 0 \text{ on } \partial D$$

Pohozaev's identity takes the form (here $F(x, u) = \int_0^u f(x, t) \, dt$)

$$\int_D [2nF(x, u) + (2 - n)uf(x, u) + 2\Sigma_{i=1}^n x_i F_{x_i}(x, u)] \, dx = \int_{\partial D} (x \cdot \nu) |\nabla u|^2 \, dS.$$

1.5.1 *Non-existence of solutions in the presence of super-critical and lower order terms*

Following K. Schmitt [164], we now derive a non-existence result, in case lower order terms are added to a super-critical term. Let $u(x)$ be a non-negative solution of the semilinear Dirichlet problem on a bounded domain $D \subset R^n$

$$\Delta u + \lambda f(u) = 0 \text{ in } D, \quad u = 0 \text{ on } \partial D, \tag{1.37}$$

with $f(u) \in C^1(\bar{R}_+)$, and λ a positive parameter. We assume that

$$f(0) = 0, \quad \text{and } f(u) > 0 \text{ for all } u > 0. \tag{1.38}$$

With $F(u) = \int_0^u f(t) \, dt$, we define, following [164],

$$M_\alpha = \sup_{u > \alpha} \frac{2n}{n-2} \frac{F(u)}{uf(u)} > 0; \tag{1.39}$$

$$m_\alpha = \max_{0 \le u \le \alpha} \frac{\frac{2n}{n-2}F(u) - M_\alpha uf(u)}{u^2}. \tag{1.40}$$

In view of (1.38), $\lim_{u \to 0} \frac{\frac{2n}{n-2}F(u) - M_\alpha uf(u)}{u^2} = (\frac{n}{n-2} - M_\alpha)f'(0)$, which implies that m_α is finite, and $m_\alpha \ge 0$. The following result is included in K. Schmitt [164], along with the references to earlier works. As before, λ_1 denotes the principal eigenvalue of $-\Delta$ on D, with zero Dirichlet condition.

Proposition 1.1. *Let D be a star-shaped domain, with respect to the origin. Assume there exists an $\alpha > 0$, such that*

$$M_\alpha = \sup_{u > \alpha} \frac{2n}{n-2} \frac{F(u)}{uf(u)} < 1 .$$

If $m_\alpha = 0$, then the problem (1.37) has no non-trivial non-negative solutions for any $\lambda > 0$. If $m_\alpha > 0$, then the problem (1.37) has no non-trivial non-negative solutions for

$$\lambda < \frac{1 - M_\alpha}{m_\alpha} \lambda_1 .$$

Proof: Writing $uf(u) = M_\alpha uf(u) + (1 - M_\alpha)uf(u)$, and dropping the positive term $\int_{\partial D}(x \cdot \nu)|\nabla u|^2 \, dS$, we obtain from Pohozaev's identity

$$\lambda \int_D \left[\frac{2n}{n-2} F(u) - M_\alpha uf(u) \right] dx \geq \lambda(1 - M_\alpha) \int_D uf(u) \, dx .$$

Write $D_1 = \{x : u(x) \leq \alpha\}$, and $D_2 = \{x : u(x) > \alpha\}$. Since $\int_{D_2} \left[\frac{2n}{n-2} F(u) - M_\alpha uf(u) \right] dx \leq 0$, we conclude

$$\lambda \int_{D_1} \left[\frac{2n}{n-2} F(u) - M_\alpha uf(u) \right] dx \geq \lambda(1 - M_\alpha) \int_D uf(u) \, dx \quad (1.41)$$
$$= (1 - M_\alpha) \int_D |\nabla u|^2 \, dx .$$

In case $m_\alpha = 0$, the integral on the left is non-positive, so that only the trivial solution is possible. If $m_\alpha > 0$, we have

$$\lambda \int_{D_1} \left[\frac{2n}{n-2} F(u) - M_\alpha uf(u) \right] dx \leq \lambda m_\alpha \int_{D_1} u^2 \, dx \leq \lambda m_\alpha \int_D u^2 \, dx .$$

Since by Poincare's inequality, $\int_D |\nabla u|^2 \, dx \geq \lambda_1 \int_D u^2 \, dx$, we obtain from (1.41),

$$\lambda m_\alpha \geq \lambda_1 (1 - M_\alpha) ,$$

and the proof follows. $\diamondsuit$

Example Let $f(u) = u^p + u^q$, with $1 < p < \frac{n+2}{n-2} < q$. For large α, $M_\alpha < 1$ and $m_\alpha > 0$. Hence, the problem (1.37) has no non-trivial non-negative solution for small $\lambda > 0$.

1.5.2 Non-existence of solutions for a class of systems

The following class of systems has attracted considerable attention recently

$$\Delta u + H_v(u, v) = 0 \text{ in } D, \quad u = 0 \text{ on } \partial D \tag{1.42}$$
$$\Delta v + H_u(u, v) = 0 \text{ in } D, \quad v = 0 \text{ on } \partial D,$$

where $H(u, v)$ is a given differentiable function, see e.g., the following surveys: D.G. de Figueiredo [55], P. Quittner and P. Souplet [154], B. Ruf [158], see also P. Korman [87]. This system is of *Hamiltonian* type, and it turns out that it has some of the properties of scalar equations. We call solution of (1.42) to be positive, if $u(x) > 0$ and $v(x) > 0$, for all $x \in D$. We consider only the classical solutions of class $C^2(D) \cap C^1(\bar{D})$. We have the following generalization of Pohozaev's identity.

Theorem 1.17. *Assume that* $H(u, v) \in C^2(R_+ \times R_+) \cap C(\bar{R}_+ \times \bar{R}_+)$. *For any positive solution of (1.42), and any real number* a, *one has*

$$\int_D [2nH(u, v) + (2 - n)(avH_v + (2 - a)uH_u)] \, dx \tag{1.43}$$
$$= 2 \int_{\partial D} (x \cdot \nu) |\nabla u| |\nabla v| \, dS.$$

Proof: Let $p = x \cdot \nabla u = \Sigma_{i=1}^n x_i u_{x_i}$, and $q = x \cdot \nabla v = \Sigma_{i=1}^n x_i v_{x_i}$. These functions satisfy the system

$$\Delta p + H_{vu}p + H_{vv}q = -2H_v \tag{1.44}$$
$$\Delta q + H_{uu}p + H_{uv}q = -2H_u.$$

We multiply the first equation in (1.42) by q, and subtract from that the first equation in (1.44) multiplied by v. The result can be written as

$$\Sigma_{i=1}^n [(u_{x_i}q - p_{x_i}v)_{x_i} + (-u_{x_i}q_{x_i} + v_{x_i}p_{x_i})] \tag{1.45}$$
$$+ H_v q - H_{vu}pv - H_{vv}qv = 2vH_v.$$

Similarly, we multiply the second equation in (1.42) by p, and subtract from that the second equation in (1.44) multiplied by u, and write the result as

$$\Sigma_{i=1}^n [(v_{x_i}p - q_{x_i}u)_{x_i} + (-v_{x_i}p_{x_i} + u_{x_i}q_{x_i})] \tag{1.46}$$
$$+ H_u p - H_{uu}pu - H_{uv}qu = 2uH_u.$$

Adding the equations (1.45) and (1.46), we put the result into the form

$$\Sigma_{i=1}^n (u_{x_i}q - p_{x_i}v + v_{x_i}p - q_{x_i}u)_{x_i} + \Sigma_{i=1}^n x_i (2H - uH_u - vH_v)_{x_i}$$
$$= 2uH_u + 2vH_v.$$

Writing,

$$\Sigma_{i=1}^n x_i \tfrac{\partial}{\partial x_i}(2H - uH_u - vH_v) = \Sigma_{i=1}^n \tfrac{\partial}{\partial x_i}\left[x_i(2H - uH_u - vH_v)\right]$$
$$-n(2H - uH_u - vH_v)\,,$$

we obtain the differential form of Pohozaev's identity

$$\Sigma_{i=1}^n \left[u_{x_i}q - p_{x_i}v + v_{x_i}p - q_{x_i}u + x_i\left(2H - uH_u - vH_v\right)\right]_{x_i}$$
$$= 2nH + (2 - n)\left(uH_u + vH_v\right)\,.$$

Integrating, we obtain as before

$$\int_D \left[2nH(u,v) + (2 - n)\left(vH_v + uH_u\right)\right]dx \tag{1.47}$$
$$= 2\int_{\partial D}(x \cdot \nu)|\nabla u||\nabla v|\,dS\,.$$

(Since we consider positive solutions, and ∂D is a level set for both u and v, we have $\nu = -\frac{\nabla u}{|\nabla u|} = -\frac{\nabla v}{|\nabla v|}$ on the boundary ∂D.) From the first equation in (1.42), $\int_D vH_v\,dx = \int_D \nabla u \cdot \nabla v\,dx$, while from the second equation $\int_D uH_u\,dx = \int_D \nabla u \cdot \nabla v\,dx$, i.e.,

$$\int_D vH_v\,dx = \int_D uH_u\,dx\,.$$

Using this in (1.47), we conclude the proof. $\diamond$

Remark Observe that by our conditions and elliptic regularity, classical solutions are in fact of class $C^3(D)$, so that all quantities in the above proof are well defined.

As a consequence, we have the following non-existence result.

Proposition 1.2. *Assume that D is a star-shaped domain with respect to the origin, and for some real constant α, and all $u > 0$, $v > 0$ we have*

$$(1 - \alpha)uH_u(u,v) + \alpha vH_v(u,v) > \frac{n}{n - 2}H(u,v)\,. \tag{1.48}$$

Then the problem (1.42) has no positive solution.

Proof: We use the identity (1.43), with $a/2 = \alpha$. Then, assuming existence of positive solution, the left hand side of (1.43) is negative, while the right hand side is non-negative, a contradiction. $\diamond$

Comparing this result to E. Mitidieri [135], observe that there is no need to require that $H_u(0,0) = H_v(0,0) = 0$.

An important subclass of (1.42) is

$$\Delta u + f(v) = 0 \text{ in } D, \quad u = 0 \text{ on } \partial D \tag{1.49}$$
$$\Delta v + g(u) = 0 \text{ in } D, \quad v = 0 \text{ on } \partial D\,,$$

which corresponds to $H(u, v) = F(v) + G(u)$, where as before, $F(v) = \int_0^v f(t)\, dt$, $G(u) = \int_0^u g(t)\, dt$. Unlike [135], we do not require that $f(0) = g(0) = 0$. Theorem 1.17 imples the following.

Theorem 1.18. *For any positive solution of (1.49), and any real number a, one has*

$$\int_D \left[2n(F(v) + G(u)) + (2 - n)\left(avf(v) + (2 - a)ug(u)\right)\right] dx \quad (1.50)$$
$$= 2 \int_{\partial D} (x \cdot \nu)|\nabla u||\nabla v|\, dS.$$

We now consider a particular system

$$\Delta u + v^p = 0 \ \text{ in } D, \ \ u = 0 \ \text{ on } \partial D \quad (1.51)$$
$$\Delta v + g(u) = 0 \ \text{ in } D, \ \ v = 0 \ \text{ on } \partial D,$$

with $g(u) \in C(\bar{R}_+)$, and a constant $p > 0$.

Theorem 1.19. *Assume that D is a star-shaped domain with respect to the origin, and*

$$nG(u) + (2 - n)\left(1 - \frac{n}{(n-2)(p+1)}\right) ug(u) < 0, \quad \text{for all } u > 0. \quad (1.52)$$

Then the problem (1.51) has no positive solution.

Proof:　We use Pohozaev's identity (1.50), with $f(v) = v^p$. We select the constant a, so that

$$2nF(v) + (2 - n)avf(v) = 0,$$

i.e., $a = \frac{2n}{(n-2)(p+1)}$. Then, assuming existence of positive solution, the left hand side of (1.50) is negative, while the right hand side is non-negative, a contradiction. $\diamond$

Observe that in case $p = 1$, the Theorem 1.5 provides a non-existence result for a biharmonic problem with Navier boundary conditions

$$\Delta^2 u = g(u) \ \text{ in } D, \ \ u = \Delta u = 0 \ \text{ on } \partial D. \quad (1.53)$$

Proposition 1.3. *Assume that D is a star-shaped domain with respect to the origin, and the condition (1.52) holds. Then the problem (1.53) has no positive solution.*

Finally, we consider the system

$$\Delta u + v^p = 0 \ \text{ in } D, \ \ u = 0 \ \text{ on } \partial D \quad (1.54)$$
$$\Delta v + u^q = 0 \ \text{ in } D, \ \ v = 0 \ \text{ on } \partial D,$$

depending on the positive parameters p and q. The curve $\frac{1}{p+1} + \frac{1}{q+1} = \frac{n-2}{n}$ is called a *critical hyperbola*. We recover the following well-known result of E. Mitidieri [135], see also R.C.A.M. Van der Vorst [186]. (Observe that we relax the restriction p, $q > 1$ from [135].)

Proposition 1.4. *Assume that p, $q > 0$, and*

$$\frac{1}{p+1} + \frac{1}{q+1} < \frac{n-2}{n}. \tag{1.55}$$

Then the problem (1.54) has no positive solution.

Proof: Condition (1.55) is equivalent to (1.52), and the Theorem 1.19 applies. $\Diamond$

In case $p = 1$, we have the following known result, see E. Mitidieri [135].

Proposition 1.5. *Assume that D is a star-shaped domain with respect to the origin, and $q > \frac{n+4}{n-4}$. Then the problem*

$$\Delta^2 u = u^q \ \text{in} \ D, \ \ u = \Delta u = 0 \ \text{on} \ \partial D \tag{1.56}$$

has no positive solution.

Since we do not require that $f(0) = g(0) = 0$, we have the following generalization for the system

$$\Delta u + v^p = 0 \ \text{in} \ D, \ \ u = 0 \ \text{on} \ \partial D \tag{1.57}$$
$$\Delta v + u^q - \alpha = 0 \ \text{in} \ D, \ \ v = 0 \ \text{on} \ \partial D.$$

Proposition 1.6. *Assume that p, $q > 0$ satisfy (1.55), and α is a non-negative constant. Then the problem (1.57) has no positive solution.*

1.5.3   Pohozhaev's identity for a version of p-Laplace equation

We consider the following version of p-Laplace equation

$$\Sigma_{i=1}^n \frac{\partial}{\partial x_i} \varphi(u_{x_i}) + f(u) = 0 \ \text{in} \ D, \ \ u = 0 \ \text{on} \ \partial D. \tag{1.58}$$

Here $\varphi(t) = t|t|^{p-1}$, with a constant $p > 1$. This is a variational equation for the functional $\int_D \left[\frac{1}{p} \left(|u_{x_1}|^p + \cdots + |u_{x_1}|^p \right) - F(u) \right] dx$. This equation is known to the experts, see P. Lindqvist [129], but it has not been studied much.

Observe that $\varphi(at) = a^{p-1}\varphi(t)$ for any constant a, and $\varphi'(t) = (p-1)|t|^{p-2}$, i.e.,

$$t\varphi'(t) = (p-1)\varphi(t). \tag{1.59}$$

Letting, as before, $z = x \cdot \nabla u = \Sigma_{i=1}^n x_i u_{x_i}$, we see that z satisfies

$$\Sigma_{i=1}^n \frac{\partial}{\partial x_i}[\varphi'(u_{x_i})z_{x_i}] + f'(u)z = -pf(u). \tag{1.60}$$

To derive (1.60), we consider $u^s \equiv u(sx)$, which satisfies

$$\Sigma_{i=1}^n \frac{\partial}{\partial x_i}\varphi(\frac{\partial}{\partial x_i}u^s) = -s^p f(u^s). \tag{1.61}$$

(To see that, it is convenient to write (1.58) as $\Sigma_{i=1}^n \varphi'(\frac{\partial}{\partial x_i}u)\frac{\partial^2}{\partial x_i^2}u + f(u) = 0$.) Then differentiating (1.61) with respect to s, and setting $s = 1$, we obtain (1.60).

Proposition 1.7. *Any solution of (1.58) satisfies*

$$\int_D [pnF(u) + (p-n)uf(u)] \, dx = (p-1)\int_{\partial D}(x{\cdot}\nu)|\nabla u|\, \Sigma_{i=1}^n\varphi(|\nabla u|\nu_i)\nu_i \, dS, \tag{1.62}$$

where ν_i is the i-th component of ν, the unit normal vector on ∂D, pointing outside.

Proof: Multiply the equation (1.58) by z, and write the result as

$$(p-1)\Sigma_{i=1}^n \frac{\partial}{\partial x_i}[z\varphi(u_{x_i})] - (p-1)\Sigma_{i=1}^n\varphi(u_{x_i})z_{x_i} + (p-1)f(u)z = 0. \tag{1.63}$$

Multiply the equation (1.60) by u, and write the result as

$$\Sigma_{i=1}^n \frac{\partial}{\partial x_i}[u\varphi'(u_{x_i})z_{x_i}] - \Sigma_{i=1}^n u_{x_i}\varphi'(u_{x_i})z_{x_i} + f'(u)uz = -puf(u). \tag{1.64}$$

We now subtract (1.64) from (1.63). In view of (1.59), we have a cancellation, and so we have

$$\Sigma_{i=1}^n \frac{\partial}{\partial x_i}\left[(p-1)z\varphi(u_{x_i}) - u\varphi'(u_{x_i})z_{x_i}\right] + [(p-1)f(u) - uf'(u)]z = puf(u).$$

As before,

$$[(p-1)f(u) - uf'(u)]z = \Sigma_{i=1}^n x_i\frac{\partial}{\partial x_i}(pF(u) - uf(u)) \tag{1.65}$$

$$= \Sigma_{i=1}^n \frac{\partial}{\partial x_i}[x_i(pF(u) - uf(u))] - n(pF(u) - uf(u)).$$

This gives us a differential form of Pohozaev's identity

$$\Sigma_{i=1}^n \frac{\partial}{\partial x_i}\left[(p-1)z\varphi(u_{x_i}) - u\varphi'(u_{x_i})z_{x_i} + x_i(pF(u) - uf(u))\right] \tag{1.66}$$

$$= pnF(u) + (p-n)uf(u).$$

Integrating, and using the divergence theorem, we conclude the proof. $\Diamond$

For star-shaped domains the right hand side of (1.62) is non-negative, so if

$$pnF(u) + (p - n)uf(u) < 0 \text{ for all } u,$$

then the problem (1.58) has no non-trivial solutions.

Example The problem

$$\Sigma_{i=1}^{n} \frac{\partial}{\partial x_i} \varphi(u_{x_i}) + u|u|^{r-1} = 0 \text{ in } D, \quad u = 0 \text{ on } \partial D$$

has no non-trivial solutions, provided the constant r satisfies $r > \frac{np-n+p}{n-p}$.

It is known that Pohozhaev's identity extends to all variational equations, see P. Pucci and J. Serrin [153], which can provide an alternative approach.

1.6   Problems at resonance

Following the publication of the classical paper of E.M. Landesman and A.C. Lazer [123], there has been an enormous amount of research into nonlinear perturbations of linear equations at resonance, of the type

$$\Delta u + \lambda_1 u + b(u) = f(x), \quad x \in D, \quad u = 0 \text{ for } x \in \partial D, \tag{1.67}$$

where D is a bounded smooth domain in R^n, and λ_1 is the principal eigenvalue of $-\Delta$ on D, with zero at the boundary condition. We shall denote by $\phi_1(x)$ the corresponding eigenfunction, normalized so that $\int_D \phi_1^2(x)\, dx = 1$. Early contributions to the subject included the classical works by A. Ambrosetti and G. Prodi [11], and M.S. Berger and E. Podolak [20]. A nice presentation and more references can be found in the book of A. Ambrosetti and G. Prodi [12]. Clearly, the problem (1.67) is not always solvable. Even in case $b(u) \equiv 0$, it is necessary (and sufficient) for existence of solution that $\int_D f(x)\phi_1(x)\, dx = 0$. It is remarkable that for some nonlinear equations one can also develop necessary and sufficient conditions for existence of solutions.

Assume first that $b(u)$ is a bounded and continuous function, which has finite limits at $\pm\infty$, and

$$b(-\infty) < b(u) < b(\infty). \tag{1.68}$$

We denote $\phi_1^\perp = \{f \in L^2(D) : \int_D f\phi_1\,dx = 0\}$, the orthogonal complement of $\phi_1(x)$ in $L^2(D)$, and decompose $f(x) = \mu_0\phi_1(x) + e(x)$, with $e(x) \in \phi_1^\perp$. Multiplying the equation (1.67) by $\phi_1(x)$, and integrating, we see that

$$b(-\infty)\int_D \phi_1(x)\,dx < \mu_0 < b(\infty)\int_D \phi_1(x)\,dx \qquad (1.69)$$

is a necessary condition for existence of solutions. The famous result of E.M. Landesman and A.C. Lazer [123] shows that it is also a sufficient condition for existence of solutions. We discuss this result next.

We introduce an extra degree of freedom, by considering the following problem: find $(\mu, u(x))$ solving

$$\Delta u + \lambda_1 u + b(u) = \mu\phi_1(x) + e(x), \quad x \in D, \quad u = 0 \text{ for } x \in \partial D. \qquad (1.70)$$

Decompose

$$u(x) = \xi\phi_1(x) + U(x), \quad U(x) \in \phi_1^\perp. \qquad (1.71)$$

We claim that for any $\xi \in R$ one can find a pair $(\mu, u(x))$ solving (1.70). Once the claim is established, the "game" is to vary ξ, to achieve $\mu = \mu_0$, and the corresponding $u(x)$ is then a solution of (1.67).

To prove the claim, we begin by plugging the ansatz (1.71) into (1.70), giving

$$\Delta U + \lambda_1 U + b(\xi\phi_1(x) + U) = \mu\phi_1(x) + e(x), \quad x \in D, \qquad (1.72)$$
$$U = 0 \text{ for } x \in \partial D.$$

Multiplying the equation (1.72) by $\phi_1(x)$, and integrating, we express

$$\mu = \int_D b(\xi\phi_1(x) + U)\phi_1\,dx. \qquad (1.73)$$

Using this in (1.72), we get

$$\Delta U + \lambda_1 U = -b(\xi\phi_1 + U) + \phi_1(x)\int_D b(\xi\phi_1 + U)\phi_1\,dx + e(x), \qquad (1.74)$$
$$U = 0 \text{ for } x \in \partial D.$$

The equations (1.73) and (1.74) provide the classical Lyapunov-Schmidt reduction, giving us the projections of the original equation on ϕ_1 and $\phi_1^\perp$ respectively. To apply a fixed point argument to (1.74), we set up a map $V \to U$, taking the subspace $\phi_1^\perp$ into itself, by solving

$$\Delta U + \lambda_1 U = -b(\xi\phi_1 + V) + \phi_1(x)\int_D b(\xi\phi_1 + V)\phi_1\,dx + e(x), \qquad (1.75)$$
$$U = 0 \text{ for } x \in \partial D.$$

The right hand side of (1.75) is orthogonal to ϕ_1, and hence (1.75) has infinitely many solutions of the form $U = U_0 + c\phi_1$, where U_0 is any solution, and c an arbitrary constant. We now fix c, so that $U \in \phi_1^\perp$. Since $b(u)$ is bounded, we have a compact map $\phi_1^\perp \to \phi_1^\perp$. By Schauder's fixed point theorem it has a fixed point, which is a solution of (1.74). We then calculate μ from (1.73), obtaining the pair $(\mu, u(x))$ solving (1.70).

Since $b(u)$ is bounded, the right hand side of (1.74) belongs to $L^p(D)$ for any $p > 1$. By the elliptic theory $U \in W^{2,p}(D)$, for any $p > 1$, and by the Sobolev imbedding

$$||U||_{L^\infty(D)} \le c, \tag{1.76}$$

for some constant c independent of ξ. Then from (1.73),

$$\mu \to b(\pm\infty) \int_D \phi_1(x) \, dx, \quad \text{as } \xi \to \pm\infty.$$

So that for any μ_0 satisfying (1.69), one expects existence of ξ, with $\mu = \mu_0$, giving us a solution of (1.67). The problem with this intuitive argument is that we do not have continuity of U in ξ (and in fact this map may be discontinuous for some values of ξ). Notice, however, that this argument implies a L^∞ a priori estimate for the problem (1.67), if μ_0 satisfies (1.69). (Indeed, we have (1.76), and we know that $|\xi|$ cannot be large, and so $u(x) = \xi\phi_1(x) + U(x)$ is bounded.) Once a priori estimate is established, existence of solutions is usually proved by degree theory. Later on we shall carry out a degree theory type argument for a system of equations.

Another way "to do business" at resonance was discovered by D.G. de Figueiredo and W.-M. Ni [56], who proved existence of solutions assuming that $b(u)$ is bounded, $ub(u) > 0$ for all $u \in R$, and the forcing term $f(x)$ has zero first harmonic, i.e., $\mu_0 = \int_\Omega f(x)\phi_1(x) \, dx = 0$. Observe that their result does not require $b(u)$ to have limits at $\pm\infty$. Their proof involved establishment of an a priori estimate, which was remarkable because such estimates usually require some conditions on $b(u)$ at infinity. In the above framework, with $u(x) = \xi\phi_1(x) + U(x)$, we have the estimate (1.76) as before, and since we consider the case $\mu_0 = 0$, (1.73) becomes

$$\int_D b(\xi\phi_1(x) + U)\phi_1 \, dx = 0.$$

If one assumes that ξ is large in absolute value and positive (negative), then in view of (1.76), the integral on the left is positive (negative), a contradiction. We conclude that $|\xi|$ is bounded, which together with (1.76) provides an a priori estimate. The proof of existence of solutions is finished by using degree theory.

Following P. Korman [102], we now extend the result of D.G. de Figueiredo and W.-M. Ni [56] to a system of two equations. The system

$$\Delta u + \lambda v + b_1(v) = f(x), \quad x \in D, \quad u = 0 \text{ for } x \in \partial D \qquad (1.77)$$

$$\Delta v + \frac{\lambda_1^2}{\lambda} u + b_2(u) = g(x), \quad x \in D, \quad v = 0 \text{ for } x \in \partial D,$$

with any $\lambda > 0$ can be seen as the case of resonance at the principal eigenvalue, similarly to the equation (1.67) (as we explain below). Similarly to [56], we assume that $b_1(u)$ and $b_2(u)$ are bounded and continuous functions, with $t b_i(t) > 0$ for all $t \in R$, $i = 1, 2$. We prove existence of solutions, provided that the first harmonics of $f(x)$ and $g(x)$ lie on a certain straight line. Let us mention that there is a considerable recent interest in systems of this type, see e.g., the recent surveys of D.G. de Figueiredo [55], and B. Ruf [158].

To explain why the system (1.77) is an analog of the resonant equation (1.67), we consider a weakly coupled linear system

$$\Delta u + \lambda v = f(x), \quad x \in D, \quad u = 0 \text{ for } x \in \partial D \qquad (1.78)$$

$$\Delta v + \bar{\lambda} u = g(x), \quad x \in D, \quad v = 0 \text{ for } x \in \partial D,$$

with given functions $f(x)$ and $g(x)$, and parameters λ and $\bar{\lambda}$. The following proposition identifies the set of non-resonant parameters λ and $\bar{\lambda}$. We denote by λ_n the eigenvalues of $-\Delta$ on D, which vanish at the boundary, and by $\phi_n(x)$ the corresponding eigenfunctions.

Proposition 1.8. *Assume that $\lambda \bar{\lambda} \neq \lambda_n^2$ for all $n \geq 1$. Then for any pair $(f(x), g(x)) \in L^2(D) \times L^2(D)$ there exists a unique solution $(u(x), v(x)) \in \left(W^{2,2}(D) \cap W_0^{1,2}(D) \right)^2$.*

Proof: Existence of solution in $L^2(D) \times L^2(D)$ follows by using the Fourier series in $\phi_n(x)$, written for u, v, f, and g, and then the standard elliptic estimates provide the extra regularity of the solution. $\diamond$

The *resonance* case is when $\lambda \bar{\lambda} = \lambda_n^2$. We shall consider the *principal* resonance case $\lambda \bar{\lambda} = \lambda_1^2$, i.e., $\bar{\lambda} = \frac{\lambda_1^2}{\lambda}$.

We now come to the main result of this section.

Theorem 1.20. *Assume that $b_1(t)$ and $b_2(t)$ are bounded and continuous functions, such that*

$$t b_i(t) > 0 \quad \text{for all } t \in R, \ i = 1, 2. \qquad (1.79)$$

Decompose $f(x) = \mu_1 \phi_1(x) + e_1(x)$, $g(x) = \nu_1 \phi_1(x) + e_2(x)$, with $e_1(x)$, $e_2(x) \in \phi_1^\perp$. Then the system (1.77) is solvable for any (μ_1, ν_1) satisfying

$$\lambda_1 \mu_1 + \lambda \nu_1 = 0 \tag{1.80}$$

with $u, v \in W^{2,p}(D) \cap W_0^{1,p}(D)$, for all $p > 2$.

The proof will be based on the following lemmas. The first one follows immediately by considering Fourier series expansion in $\phi_n(x)$.

Lemma 1.6. *The solution set of the linear system*

$$\Delta u + \lambda v = 0, \quad x \in D, \quad u = 0 \ \text{for} \ x \in \partial D \tag{1.81}$$
$$\Delta v + \tfrac{\lambda_1^2}{\lambda} u = 0, \quad x \in D, \quad v = 0 \ \text{for} \ x \in \partial D$$

is $(u, v) = c(\phi_1, \tfrac{\lambda_1}{\lambda} \phi_1)$, where c is an arbitrary constant. In particular, the only solution of (1.81) in $\phi_1^\perp \times \phi_1^\perp$ is $(0, 0)$.

Lemma 1.7. *Let U, $V \in \phi_1^\perp$ be solutions of*

$$\Delta U + \lambda V = f(x), \quad x \in D, \quad u = 0 \ \text{for} \ x \in \partial D \tag{1.82}$$
$$\Delta V + \tfrac{\lambda_1^2}{\lambda} U = g(x), \quad x \in D, \quad v = 0 \ \text{for} \ x \in \partial D,$$

with $f(x)$, $g(x) \in L^\infty(D)$. Then for any $p > 1$ one can find a constant $c > 0$, such that

$$||U||_{W^{2,p}(D)} + ||V||_{W^{2,p}(D)} \le c(||f||_{L^p(D)} + ||g||_{L^p(D)}). \tag{1.83}$$

By the Sobolev imbedding this implies that $||U||_{L^\infty(D)} + ||V||_{L^\infty(D)} \le c_1$, for some constant $c_1 > 0$.

Proof: Standard elliptic estimates imply that

$$||U||_{W^{2,p}} + ||V||_{W^{2,p}} \le c(||f||_{L^p} + ||g||_{L^p} + ||U||_{L^p} + ||V||_{L^p}).$$

The estimate (1.83) will follow, once we prove that

$$||U||_{L^p} + ||V||_{L^p} \le c(||f||_{L^p} + ||g||_{L^p}). \tag{1.84}$$

Assume for definiteness that $||f||_{L^p} \ge ||g||_{L^p}$. Dividing both equations in (1.82) by the same constant $||f||_{L^p}$, and redefining U and V, we may assume that $||f||_{L^p} = 1$ and $|g||_{L^p} \le 1$. Assuming that the estimate (1.84) is not possible with any constant c, we could find a sequence $\{f_n, g_n\}$, with $||f_n||_{L^p} = 1$ and $|g_n||_{L^p} \le 1$, and the corresponding solutions of (1.82) $\{U_n, V_n\} \in \phi_1^\perp \times \phi_1^\perp$, so that

$$||U_n||_{L^p} + ||V_n||_{L^p} \ge n(1 + ||g_n||_{L^p}).$$

In particular, $||U_n||_{L^p} + ||V_n||_{L^p} \to \infty$, as $n \to \infty$. Define $u_n = \dfrac{U_n}{||U_n||_{L^p} + ||V_n||_{L^p}}$ and $v_n = \dfrac{V_n}{||U_n||_{L^p} + ||V_n||_{L^p}}$. They satisfy

$$\Delta u_n + \lambda v_n = \frac{f_n(x)}{||U_n||_{L^p} + ||V_n||_{L^p}} \tag{1.85}$$

$$\Delta v_n + \frac{\lambda_1^2}{\lambda} u_n = \frac{g_n(x)}{||U_n||_{L^p} + ||V_n||_{L^p}}.$$

Since $||u_n||_{L^p} < 1$, $||v_n||_{L^p} < 1$, we get uniform in n bounds for $||u_n||_{W^{2,p}}$ and $||v_n||_{W^{2,p}}$. In a standard way, along a subsequence $\{u_n, v_n\} \to (u, v) \in \phi_1^\perp \times \phi_1^\perp$, with (u, v) solving (1.81). Hence $u = v = 0$ by Lemma 1.6, but $||u + v||_{L^p} = 1$, a contradiction. $\diamondsuit$

The following lemma provides the crucial a priori estimate. As was the case with the result of D.G. de Figueiredo and W.-M. Ni, it is remarkable that this estimate does not require any conditions on $b_i(t)$ at infinity (which are usually needed to get a priori estimates).

Lemma 1.8. *In the conditions of the Theorem 1.20, there is a constant $c > 0$, so that any solution of (1.77) satisfies*

$$||u||_{L^2(\Omega)} + ||v||_{L^2(\Omega)} \le c.$$

Proof: Decompose $u(x) = \xi_1 \phi_1(x) + U(x)$, $v(x) = \eta_1 \phi_1(x) + V(x)$, with $U(x), V(x) \in \phi_1^\perp$. The system (1.77) becomes

$$\Delta U + \lambda V + (-\lambda_1 \xi_1 + \lambda \eta_1)\phi_1 + b_1(\eta_1 \phi_1 + V) = \mu_1 \phi_1 + e_1, \quad (1.86)$$

$$\Delta U + \frac{\lambda_1^2}{\lambda} U + \frac{\lambda_1}{\lambda}(\lambda_1 \xi_1 - \lambda \eta_1)\phi_1 + b_2(\xi_1 \phi_1 + U) = \nu_1 \phi_1 + e_2.$$

We claim that

$$|-\lambda_1 \xi_1 + \lambda \eta_1| \le c, \tag{1.87}$$

for some constant $c > 0$. Indeed, multiply the first equation in (1.86) by ϕ_1, and integrate over D. Since $\int_D \Delta U \phi_1 \, dx = \int_D V \phi_1 \, dx = 0$, while b_1 is a bounded function, the claim follows. By Lemma 1.7 it follows that

$$||U||_{L^\infty(D)} + ||V||_{L^\infty(D)} \le c_1, \tag{1.88}$$

for some constant $c_1 > 0$.

To complete the proof, we need an a priori estimate of the first harmonics ξ_1 and η_1. By (1.87), if either one of ξ_1 and η_1 is large and positive (negative), so is the other one. Assume, for definiteness, that ξ_1 and η_1 are both negative, and large in absolute value. Multiply the first equation in (1.86) by $\lambda_1 \phi_1$, the second one by $\lambda \phi_1$, integrate over D, and add the results. By our condition (1.80)

$$0 = \lambda_1 \mu_1 + \lambda \nu_1 = \int_D \left[\lambda_1 b_1 (\eta_1 \phi_1(x) + V(x)) + \lambda b_2 (\xi_1 \phi_1(x) + U(x)) \right] \phi_1(x) \, dx \, .$$

We claim that the integral on the right is negative, which gives us a contradiction. Indeed, by (1.88), $\eta_1 \phi_1(x) + V(x) < 0$ and $\xi_1 \phi_1(x) + U(x) < 0$ over D, and then by our conditions the functions b_1 and b_2 are negative. $\diamond$

We are now ready to finish the proof.

Proof of the Theorem 1.20 Letting $w = (u, w)$, we rewrite the system (1.77) in the operator form

$$w = T(w) \, ,$$

where (we abbreviate $b_1 = b_1(v)$ and $b_2 = b_2(u)$)

$$T(w) = \left(\Delta^{-1} \left(-\lambda v - b_1 + \mu_1 \phi_1 + e_1 \right), \Delta^{-1} \left(-\frac{\lambda_1^2}{\lambda} u - b_2 + \nu_1 \phi_1 + e_2 \right) \right).$$

T is a compact map $L^2(D) \times L^2(D) \to L^2(D) \times L^2(D)$. We denote $\mathbf{L}^2 = L^2(D) \times L^2(D)$, with the norm $||w||_{\mathbf{L}^2}^2 = ||u||_{L^2(D)}^2 + ||v||_{L^2(D)}^2$. Following D.G. de Figueiredo and W.-M. Ni [56], define the operator

$$T_k(w) = \frac{1}{k+1} T(w) - \frac{k}{k+1} T(-w), \quad 0 \leq k \leq 1 \, ,$$

which is compact for all k, $T_0 = T$, and T_1 is an odd operator. It is known, see e.g., L. Nirenberg [139], that the Leray-Schauder degree

$$deg(I - T_1, B_R, 0) \neq 0$$

for any ball $B_R = \{ w \in \mathbf{L}^2 : ||w||_{\mathbf{L}^2} \leq R \}$. We claim that there is an R such that

$$w - T_k(w) \neq 0, \quad \text{for } ||w||_{\mathbf{L}^2} \leq R, \ 0 \leq k \leq 1 \, .$$

Then by the homotopy invariance of the degree, $deg(I - T, B_R, 0) \neq 0$, which implies that the system (1.77) has a solution. To prove the claim, we need a uniform in k a priori bound for the equation

$$w - T_k(w) = 0 \, ,$$

which is equivalent to

$$\Delta u + \lambda v + \frac{1}{k+1} b_1(v) - \frac{k}{k+1} b_1(-v) = \frac{1-k}{1+k} \left(\mu_1 \phi_1 + e_1 \right) \qquad (1.89)$$

$$\Delta v + \frac{\lambda_1^2}{\lambda} u + \frac{1}{k+1} b_2(u) - \frac{k}{k+1} b_2(-u) = \frac{1-k}{1+k} \left(\nu_1 \phi_1 + e_2 \right) \, .$$

Clearly, the condition (1.80) on the first harmonics is satisfied for all k. Letting $b_i^k(t) = \frac{1}{k+1} b_i(t) - \frac{k}{k+1} b_i(-t)$, $i = 1, 2$, we see that these functions are uniformly bounded in k, satisfying the condition (1.79). By Lemma 1.8, we conclude a uniform in k a priori bound for solutions of (1.89), concluding the proof. $\diamond$

Chapter 2

Curves of Solutions on Balls

2.1 Preliminary results

The story begins with the classical theorem of B. Gidas, W.-M. Ni and L. Nirenberg [63], which states that in case the domain D is a ball in R^n, say a unit ball, then any positive solution of the semilinear Dirichlet problem (1.1) from the preceding chapter is necessarily radially symmetric, i.e., $u = u(r)$, with $r = |x|$, and so the problem turns into an ODE

$$u''(r) + \frac{n-1}{r}u'(r) + \lambda f(u(r)) = 0 \text{ for } 0 < r < 1, \quad u'(0) = u(1) = 0. \quad (2.1)$$

Moreover, the theorem asserts that

$$u'(r) < 0, \quad \text{for all } 0 < r < 1. \quad (2.2)$$

In particular, this implies that $u(0)$ gives the maximum value of solution. The only assumption of this remarkable theorem is a slight smoothness assumption on $f(u)$, which is considerably weaker than the standing assumption of this chapter:

$$f(u) \in C^2(\bar{R}_+). \quad (2.3)$$

As we had mentioned above, C.S. Lin and W.-M. Ni [127] proved that solutions of the corresponding linearized problem are also radially symmetric, i.e., they satisfy

$$w''(r) + \frac{n-1}{r}w'(r) + \lambda f'(u(r))w(r) = 0 \text{ for } 0 < r < 1, \quad (2.4)$$
$$w'(0) = w(1) = 0.$$

On the surface, the equation (2.1) appears to be singular at $r = 0$. In reality there is no singularity at $r = 0$, after all this difficulty is introduced only by the spherical coordinates. In fact, one can easily compute all of the derivatives $u^{(k)}(0)$, see P. Korman [95]. One can write down the Maclaurin series of solution, and show that it converges for small r, provided that $f(u)$ is analytic, see [95]. Since in this chapter we only assume that $f(u) \in C^2(\bar{R}_+)$, let us recall the following result.

Lemma 2.1. *For any $\alpha > 0$, consider the initial value problem*

$$u''(r) + \tfrac{n-1}{r}u'(r) + \lambda f(u(r)) = 0 \ for \ 0 < r < 1, \qquad (2.5)$$
$$u(0) = \alpha, \ u'(0) = 0 \,.$$

Then one can find an $\epsilon > 0$, so that this problem has a unique solution on the interval $[0, \epsilon)$.

The proof can be found in L.A. Peletier and J. Serrin [148], see also J.A. Iaia [74], or P. Quittner and P. Souplet [154].

We show next that the maximum value of solution $u(0) = \alpha$ can be used as a global parameter, i.e., it is impossible for two solutions of (2.1) to share the same α. This result has been known for a while, see e.g., E.N. Dancer [47].

Lemma 2.2. *The value of $u(0) = \alpha$ uniquely identifies the solution pair $(\lambda, u(r))$ (i.e., there is at most one λ, with at most one solution $u(r)$, so that $u(0) = \alpha$).*

Proof: Assume, on the contrary, that we have two solution pairs $(\lambda, u(r))$ and $(\mu, v(r))$, with $u(0) = v(0) = \alpha$. Clearly, $\lambda \neq \mu$, since otherwise we have a contradiction with uniqueness of initial value problems guaranteed by Lemma 2.1. (Recall that $u'(0) = v'(0) = 0$.) Then $u(\tfrac{1}{\sqrt{\lambda}}r)$ and $v(\tfrac{1}{\sqrt{\mu}}r)$ are both solutions of the same initial value problem

$$u'' + \frac{n-1}{r}u'(r) + f(u) = 0, \quad u(0) = \alpha, \ u'(0) = 0 \,, \qquad (2.6)$$

and hence $u(\tfrac{1}{\sqrt{\lambda}}r) = v(\tfrac{1}{\sqrt{\mu}}r)$, but that is impossible, since the first function has its first root at $r = \sqrt{\lambda}$, while the second one at $r = \sqrt{\mu}$. $\Diamond$

In addition to its theoretical significance, this lemma is the key to numerical computation of bifurcation diagrams. For a sequence of values $\alpha = \alpha_i$ we compute the first root $r = r_i$ for the solution of (2.6). (The initial value problem (2.6) is easy to solve numerically, say by using the Runge-Kutta method.) Then $\lambda_i = r_i^2$ is the corresponding value of λ in (2.1). We then plot all the points (λ_i, α_i), obtaining the solution curve.

According to the lemma above, these two-dimensional curves give us a faithful picture of the solution set of the nonlinear PDE (1.1).

We shall need the following consequence of the Sturm Comparison Theorem.

Lemma 2.3. *Let $u(r, \lambda)$ be a positive solution of (2.1), that exists for all $\lambda > \bar{\lambda}$, for some $\bar{\lambda} > 0$. Assume that*

$$\lim_{u \to \infty} \frac{f(u)}{u} = \infty.$$

Then, as $\lambda \to \infty$, $u(r, \lambda)$ does not tend to infinity over any subinterval of $(0, 1)$.

Proof: Assume, on the contrary, that $u(r, \lambda) \to \infty$ for all $r \in [0, r_1)$, for some fixed $r_1 > 0$ (keep in mind that $u(r)$ is a decreasing function). Then for any constant $M > 0$ we can find a λ_0, so that for all $\lambda \geq \lambda_0$

$$\lambda \frac{f(u)}{u} \geq M, \quad \text{for all } r \in [0, r_1). \tag{2.7}$$

We now write our equation (2.1) in the form

$$u''(r) + \frac{n-1}{r} u'(r) + \lambda \frac{f(u)}{u} u = 0,$$

and compare it with

$$v''(r) + \frac{n-1}{r} v'(r) + M u = 0. \tag{2.8}$$

Changing variables $v = z r^{\frac{2-n}{2}}$, we reduce (2.8) to a Bessel's equation, and so it has a solution $v = r^{\frac{2-n}{2}} J_{\frac{n-2}{2}}(\sqrt{M} r)$, which has a zero on $(0, r_1)$, provided that M is large. By Sturm's comparison theorem, the same must be true for $u(r) > 0$, a contradiction, proving that the length of the interval on which solution gets large is decreasing, as $\lambda \to \infty$. $\diamond$

The following result from P. Korman [84] gives a detailed description of the solution shape for large λ. If, for some reason, solutions cannot be of that shape, it follows that there are no positive solutions of (2.1) for large λ. Recall that a root z of $f(u)$ is called *stable* if $f(z) = 0$ and $f'(z) < 0$, it is *unstable* if $f(z) = 0$ and $f'(z) > 0$.

Theorem 2.1. *Assume that all non-negative roots of $f(u)$ are either stable or unstable. Let $u(r, \lambda)$ be a positive solution of (2.1), that exists for all $\lambda > \bar{\lambda}$, for some $\bar{\lambda} > 0$. Assume that $u(r, \lambda)$ does not tend to infinity over any subinterval of $(0, 1)$ (which will happen if e.g., $\lim_{u \to \infty} \frac{f(u)}{u} = \infty$, or* *if there is $u_0 > 0$, so that $f(u) \leq 0$ for $u \geq u_0$). Then the interval $[0, 1)$ can be decomposed into a union of open intervals, whose total length $= 1$, so that on each such subinterval $u(r, \lambda)$ tends to a stable root of $f(u)$, as $\lambda \to \infty$.*

Proof: Since $u(r, \lambda)$ does not tend to infinity over any subinterval of $(0, 1)$, it follows that for any $\delta > 0$ we can find a constant $M_0 > 0$, so that

$$u(r, \lambda) < M_0 \text{ on } (\delta, 1), \quad \text{for all } \lambda \geq \lambda_0. \tag{2.9}$$

We claim that on any subinterval of $(\delta, 1)$, the solution $u(r, \lambda)$ must tend to a root of $f(u)$. Indeed, if $u(r, \lambda)$ failed to tend to a root of $f(u)$ on some subinterval $I \in (\delta, 1)$, then from the equation (2.1) we see that $r^{n-1} u'$ would have to become large in absolute value over I. This would imply that $u(r)$ has a large variation on $I \in (\delta, 1)$, contradicting the fact that the variation of $u(r)$ on $(\delta, 1)$ is less than M_0, by (2.9). Hence, the values of $u(r, \lambda)$ have to accumulate at roots of $f(u)$, as $\lambda \to \infty$.

We show next that solution has to accumulate at stable roots of $f(u)$. We outline the argument first. Throughout the argument one should keep in mind that $u(r)$ is strictly decreasing. We begin by showing that solution cannot tend to an unstable root from below, over some subinterval. Indeed, below of an unstable root solution is convex, and it remains convex until it drops down to a preceding root (or zero, if there is no preceding root). Convex and decreasing function, with a fixed drop, cannot tend to a constant. If one assumes that solution tends to an unstable root from above, over some subinterval, then we show that derivative must be large in absolute value at the right end of that interval. To rectify the function $u(r)$ over that interval, $u''(r)$ must be negative and large in absolute value, from which similarly to the above, we conclude large variation of the solution, contradicting (2.9). We present the details next.

Assume, on the contrary, that there is an interval $J \equiv (r_1, r_2) \subset (\delta, 1)$, so that

$$u(r, \lambda) \to \bar{u} \text{ over } J, \text{ as } \lambda \to \infty, \tag{2.10}$$

and $\bar{u}$ is an unstable root of $f(u)$, i.e., $f(u)$ is negative to the left of $\bar{u}$, and positive to the right. Denote by $\xi = \xi(\lambda)$ the point where $u(\xi) = \bar{u}$. We claim that there is a $\bar{r} \in (r_1, r_2)$, so that ξ will be excluded from $(r_1, \bar{r})$, as λ gets large. Indeed, we see from the equation (2.1) that $u(r)$ is convex just below the value of $\bar{u}$, and it remains convex for increasing r, at least until $u(r)$ will drop below the root of $f(u)$ which precedes $\bar{u}$, or zero (so that $f(u)$

can change to being positive). But (2.10) implies that $u(r)$ would have to change to being concave at values just a little below $\bar{u}$, a contradiction. (A convex and decreasing function, with a fixed drop, cannot tend from below to a constant over a subinterval.)

Let again ξ be such that $u(\xi) = \bar{u}$. By above, $\xi \in (\bar{r}, r_2)$ for large λ. We claim that $|u'(\xi)|$ is large, for large λ. Indeed, denote by $\underline{u}$ the root of $f(u)$ preceding $\bar{u}$, and define $\underline{u} = 0$ in case $f(u) < 0$ on $[0, \bar{u})$. Define η by $u(\eta) = \underline{u}$, and $0 < c_0 \equiv -\int_{\underline{u}}^{\bar{u}} f(u)\,du$. Multiplying the equation (2.1) by $r^{n-1}u'$, and integrating over (ξ, η), we get

$$\tfrac{1}{2}\xi^2 u'^2(\xi) = \tfrac{1}{2}\eta^2 u'^2(\eta) + \lambda \int_\xi^\eta r^{2(n-1)}f(u)u'\,dr$$
$$\geq \lambda \xi^{2(n-1)} \int_\xi^\eta f(u)u'\,dr = \lambda \xi^{2(n-1)}c_0,$$

which implies that $|u'(\xi)| \geq c_1\sqrt{\lambda}$ for large λ, with some $c_1 > 0$, proving the claim.

When we decrease r from ξ toward the interval $(r_1, \bar{r})$, we need $u''(r)$ to be large in absolute value and negative in order to decrease $|u'(r)|$. (The value of $|u'(r)|$ must be small over the interval $(r_1, \bar{r})$ for large λ, since $u(r)$ tends to a constant over that interval.) From the equation (2.1), $u'' \geq -\lambda f(u)$, i.e., $|u''| = -u'' \leq \lambda f(u)$, so that $\lambda f(u)$ must get large somewhere on $(\bar{r}, \xi)$. But as we decrease r, we increase $u(r)$, so the quantity $\lambda f(u)$ once it gets large can only further increase. It follows from the equation (2.1) that $\left(r^{n-1}u'\right)'$ is large over $(r_1, \bar{r})$. This implies a large variation of $u(r)$ over this interval, a contradiction, finishing the proof. $\diamondsuit$

Example Assume that $f(u) < 0$ for $0 < u < \bar{u}$, with some $\bar{u} > 0$, $f(u) > 0$ for $u > \bar{u}$, and $\lim_{u\to\infty} \frac{f(u)}{u} = \infty$, e.g., $f(u) = u^p - 1$, with $p > 1$. Then the problem

$$u'' + \frac{n-1}{r}u'(r) + \lambda f(u) = 0, \quad \text{for } 0 < r < 1, \quad u'(0) = u(1) = 0$$

has no positive solution for λ large enough. Indeed, since $f(u)$ has no stable roots, solution cannot exhibit the behavior described in the above theorem, and hence the solution cannot exist for all large λ.

Example The function $f(u) = u^2 - u^3$ has a stable root $u = 1$, and the root $u = 0$, which is neither stable or unstable. We shall see later on that the solution curve of

$$u'' + \frac{n-1}{r}u'(r) + \lambda(u^2 - u^3) = 0, \quad \text{for } 0 < r < 1, \quad u'(0) = u(1) = 0$$

has two branches, with the upper branch tending to 1, and the lower one tending to 0.

A word on notation. We shall denote derivatives of $u(r)$ by either $u'(r)$ or $u_r(r)$, and mix both notations when it helps to make proofs more transparent.

We shall be mostly looking for positive solutions of (2.1). One reason for this emphasis, is that only positive solutions have a chance to be stable, a property significant for applications. Let us recall the notion of stability. For any solution $u(r)$ of (2.1) let $(\mu_1, w(r))$ denote the principal eigenpair of the corresponding linearized equation, i.e., $w(r) > 0$ satisfies

$$w''(r) + \tfrac{n-1}{r} w'(r) + \lambda f'(u) w + \mu_1 w(r) = 0 \ \text{ for } 0 \le r < 1, \quad (2.11)$$
$$w'(0) = w(1) = 0 \,.$$

The solution $u(x)$ of (2.1) is called *unstable* if $\mu_1 < 0$, it is *stable* if $\mu_1 > 0$, and *neutrally stable* in case $\mu_1 = 0$. (This is so called linear stability, which means, in particular, that solutions of the corresponding heat equation, with the initial data near $u(r)$ will tend to $u(r)$, as $t \to \infty$, see e.g., the book by D. Henry [70].) The following result is probably known (a very similar argument was used in C.S. Lin and W.-M. Ni [127]).

Theorem 2.2. *Let $u(r)$ be a solution of (2.1) that changes sign on $(0,1)$. Then $u(r)$ is unstable.*

Proof: Let $(\mu_1, w(r))$ denote the principal eigenpair of (2.11). Assume on the contrary that $\mu_1 \ge 0$. Since $u(r)$ changes sign, it has either a point of minimum followed by a point of maximum, or the other way around. Assume it is the former case (the other case is similar), i.e., there are points $0 < r_1 < r_2 < 1$, such that $u'(r_1) = u'(r_2) = 0$ and $u'(r) > 0$ on (r_1, r_2). Differentiate (2.1)

$$u_r''(r) + \frac{n-1}{r} u_r'(r) + \lambda f'(u(r)) u_r - \frac{n-1}{r^2} u' = 0. \quad (2.12)$$

Observe that $u''(r_1) > 0$ and $u''(r_2) < 0$ (Clearly, $u''(r_1) \ge 0$; $u'(r)$ satisfies a linear equation (2.12), so it cannot vanish together with its derivative at r_1, since it is not identically zero.) Denoting $p(r) = r^{n-1}(u''(r)w(r) - u'(r)w'(r))$, we have by combining the equations (2.11) and (2.12)

$$p'(r) = \mu_1 u'(r) w(r) + \frac{n-1}{r^2} u'(r) w(r) \ge 0, \quad \text{for } r \in (r_1, r_2).$$

We see that $p(r)$ is non-decreasing on (r_1, r_2). On the other hand, $p(r_1) = r_1^{n-1} u''(r_1) w(r_1) > 0$ and $p(r_2) = r_2^{n-1} u''(r_2) w(r_2) < 0$, a contradiction. $\diamondsuit$

In case $n = 1$, this result was proved by R. Schaaf [161], see also P. Korman [92].

We now recall the general scheme for proving uniqueness and exact multiplicity results, introduced in P. Korman, Y. Li and T. Ouyang [108] and [109], and developed further in T. Ouyang and J. Shi [144], [145], and in a number of other papers, see e.g., [51], [92]. It consists roughly of five steps.

1. Show existence of positive solutions of (2.1) at some λ. By the Theorem 1.2, this implies existence of a curve of positive solutions.

2. Show that any non-trivial solution $w(r)$ of the linearized problem (2.4) can be chosen to be positive.

3. Using Theorem 1.3, and Step 2, show that only turns to the left, or only turns to the right, are possible on the solution curve. It follows that there is at most one turn.

4. Using Lemma 2.2, and monotonicity of one of the branches, show uniqueness of the solution curve.

5. Study the behavior of the unique solution curve, as $u(0) \to 0$, and as $u(0) \to \infty$.

Steps 2 and 3 require the most effort. As a broad overview, one could say that the Step 2 is tough, but manageable, while the Step 3 is where the main bottle-neck is at present. The only cases where the direction of the turn can be computed are the following: $f(u)$ is convex, $f(u)$ is concave, $f(u)$ is concave-convex (i.e., concave for $u \in (0, u_0)$, convex for $u > u_0$, for some u_0) and $f(0) \geq 0$, and finally $f(u)$ is convex-concave and $f(0) \leq 0$.

2.2 Positivity of solution to the linearized problem

After appearance of the classical paper of B. Gidas, W.-M. Ni and L. Nirenberg [63], the study of semilinear equations on balls began in earnest. (Of course, there were a number of results on radial solutions before, most notably another classic D.D. Joseph and T.S. Lundgren [75], but now radial solutions represented *all* solutions of the original PDE.) It soon became clear that it is hard to prove uniqueness of solutions, even for relatively simple nonlinearities $f(u)$, like $f(u) = -u + u^p$, $f(u) = u + u^p$, and even the case $f(u) = u^p$ was non-trivial (it was covered in [63]). After a great effort by a number of people, uniqueness question for these equations was settled by the early nineties. We mention the following contributions: W.-M. Ni [137], W.-M. Ni and R.D. Nussbaum [138], M.K. Kwong [120], M.K. Kwong and L. Zhang [122], L. Zhang [195], P. N. Srikanth [178], and Adimurthi and S. L. Yadava [3]. In particular, these papers have developed the method of test functions, which relies on a version of the Sturm Comparison Theorem. The proof of positivity of $w(r)$ in P. Korman, Y. Li and T. Ouyang [109], as well as in T. Ouyang and J. Shi [144], [145], depended on the methods developed in the above mentioned papers. Here we present another approach, taken from P. Korman and T. Ouyang [116], which is based on a paper of M. Tang [182].

While there are several general methods for proving positivity of solution for the linearized problem, their application depends on the specific form of $f(u)$. We begin with $f(u)$ behaving like a cubic. We assume that $f \in C^2(\bar{R}_+)$ satisfies $f(0) \leq 0$, and

$$f(u) < 0, \text{ for } u \in (0, b), \text{ and } f(u) > 0, \text{ for } u \in (b, c), \qquad (2.13)$$

for some numbers $0 < b < c \leq \infty$. We assume that $u(r)$ is a positive solution of (2.1), which is singular, i.e., the linearized problem (2.4) admits a non-trivial solution $w(r)$. We shall need the following functions, introduced by M. Tang [182]

$$\xi(r) = r^{n-1} \left(u'w - uw' \right),$$

$$\zeta(r) = r^n \left[u'w' + f(u)w \right] + (n-2)r^{n-1}u'w.$$

As was observed in M. Tang [182], these functions have the following derivatives

$$\xi'(r) = r^{n-1} \left[uf'(u) - f(u) \right] w, \qquad (2.14)$$

$$\zeta'(r) = 2f(u)wr^{n-1}. \qquad (2.15)$$

If θ is defined by the relation $\int_0^\theta f(t)\, dt = 0$, it follows by the Theorem 1.9 that

$$u(0) > \theta. \qquad (2.16)$$

We wish to show that $w(r)$ cannot vanish. As a first step, we show that it cannot vanish close to $r = 1$.

Lemma 2.4. *Let η be such that $u(\eta) = b$. Then $w(r)$ cannot vanish on the interval $[\eta, 1)$ (where $u(r) \leq b$, and $f(u) \leq 0$).*

Proof: Assuming the contrary, let ξ denote the largest root of $w(r)$ on $[\eta, 1)$. Integrate (2.15) over the interval $(\xi, 1)$

$$u'(1)w'(1) - \xi^n u'(\xi)w'(\xi) = 2 \int_{\xi}^{1} f(u)wr^{n-1}\, dr\,.$$

If we assume for definiteness that $w'(1) < 0$, and $w'(\xi) > 0$, then the quantity on the left is non-negative, while the integral on the right is negative, a contradiction. $\diamond$

We shall also need the following function from M. Tang [182]:

$$Q(r) = r^n \left[u'^2 + f(u)u \right] + (n-2)r^{n-1}u'u\,.$$

The following result of P. Korman and T. Ouyang [116] was inspired by M. Tang [182].

Theorem 2.3. *([116]) Let $u(r)$ be a positive singular solution of (2.1), and $w(r)$ a non-trivial solution of the linearized problem (2.4). Recall that η is such that $u(\eta) = b$.*
(i) *Assume that*

$$K(u) \equiv \frac{uf'(u)}{f(u)} \ \text{is decreasing for } u \in (b,c), \text{ and } \lim_{u \to c} K(u) > 1; \quad (2.17)$$

$$Q(r) > 0 \ \text{ for } r \in (0,\eta). \quad (2.18)$$

Then $w(r)$ cannot have more than one interior root on $(0,\eta)$.

(ii) *If, in addition,*

$$K(u) < 1 \ \text{ for } u \in (0,b), \quad (2.19)$$

then $w(r)$ has no interior roots on $(0,1)$, i.e., we may assume that $w(r) > 0$ on $[0,1)$.

Proof: (i) Assume, on the contrary, that $w(r)$ has two interior roots $0 < \tau < \bar{\tau} < \eta$, so that $w(r) > 0$ on $[0,\tau)$ and $w(r) < 0$ on $(\tau,\bar{\tau})$. By Lemma 2.4, $f(u(r)) > 0$ on $[0,\bar{\tau}]$. We consider a function $O(r) = \frac{2}{\gamma'}\xi(r) - \zeta(r)$, with a positive constant γ' to be specified. Using (2.14) and (2.15), we compute

$$O'(r) = \frac{2}{\gamma'} f(u(r))r^{n-1} \left(K(u(r)) - \gamma \right) w(r), \quad (2.20)$$

with $\gamma = \gamma' + 1$. Observe that by our conditions, $K(u(r))$ is an increasing function of r, with $K(u(0)) > 1$. We now fix γ (and hence γ') by setting

$\gamma = K(u(\tau))$. Then, we see from (2.20) that the function $O(r)$ is decreasing on $(0, \bar{\tau})$. (Observe that at $r = \tau$, both $K(u(r)) - \gamma$ and $w(r)$ change sign.) Since $O(0) = 0$, it follows that

$$O(r) < 0 \text{ for } r \in (0, \bar{\tau}).$$

Since $\xi(\tau) > 0$, while $\xi(\bar{\tau}) < 0$, there exists a $t \in (\tau, \bar{\tau})$ so that $\xi(t) = 0$. It follows that

$$\zeta(t) = -O(t) > 0. \tag{2.21}$$

On the other hand, since $\frac{u(t)}{w(t)} = \frac{u'(t)}{w'(t)}$, and using (2.18)

$$\zeta(t) = \left[t^n \left(u'w'\frac{u}{w} + f(u)u \right) + (n-2)t^{n-1}u'u \right] \frac{w}{u} = Q(t)\frac{w(t)}{u(t)} < 0,$$

contradicting (2.21).

(ii) We now exclude the possibility that $w(r)$ has exactly one interior root τ, so that $w(r) > 0$ on $[0, \tau)$ and $w(r) < 0$ on $(\tau, 1)$. The argument is similar. Clearly, $\tau \in (0, \eta)$. Again, we consider the function $O(r)$, and set $\gamma = K(u(\tau)) > 1$. Thanks to the assumption (2.19), we see from (2.20) that $O(r)$ is decreasing on $(0, 1)$. (Observe that at $r = \tau$, both $K(u(r)) - \gamma$ and $w(r)$ change sign, while at $r = \eta$, both $K(u(r)) - \gamma$ and $f(u(r))$ change sign.) It follows that $O(1) < 0$, i.e., $\zeta(1) = -O(1) > 0$. But $\zeta(1) = u'(1)w'(1) \leq 0$, a contradiction (it is possible to have $u'(1) = 0$, when $f(0) < 0$). $\diamondsuit$

Remarks

(1) Assume that we know that $w(r)$ cannot vanish in the region where $u(r) > \rho$, for some $\rho \in (b, c)$. Then we may replace the condition $\lim_{u \to c} K(u) > 1$, by requiring that $K(\rho) > 1$ and $K(u) < K(\rho)$, for $u \in (\rho, c)$.

(2) The first part of the theorem was also proved in T. Ouyang and J. Shi [145] by a different and more involved method. The second part could be also proved by using a certain integral identity of M.K. Kwong and L. Zhang [122], see [109] or [145], and Theorem 2.21 below, for details. We see that the proof above provides a unified and simplified approach.

(3) We needed positivity of $Q(r)$ only between the roots of $w(r)$. So if, for example, $w(r)$ cannot vanish in the region where $u(r)$ is greater than some constant, then we do not need $Q(r)$ to be positive in that region. In particular, in case $n = 2$ the condition (2.18) can be dropped, since then $Q(r) > 0$ for $r \in [0, \eta)$.

(4) We now show that are many $f(u)$ for which $K(u)$ is decreasing. To emphasize the dependence on $f(u)$, we denote $K_f(u) = \frac{uf'(u)}{f(u)}$. Then we have for any two functions f and g

$$K_f = K_g + K_{\frac{f}{g}}. \tag{2.22}$$

So if K_f is decreasing, the same is true for $K_{u^p f}$, for any $p > 0$. Also, we see from (2.22) that

$$K_{f^2} = K_f + F_{\frac{f^2}{f}} = 2K_f,$$

and, in general, $K_{f^m} = mK_f$. Hence, K_{f^m} is decreasing, provided that K_f is decreasing.

We now give some conditions on $f(u)$ under which $Q(r) > 0$, i.e., the condition (2.18) holds. We let

$$P(r) = r^n \left(u'^2 + 2F(u) \right) + (n-2)r^{n-1}u'u,$$

where, as usual, $F(u) = \int_0^u f(t)\,dt$. Then

$$Q(r) - P(r) = r^n \left(uf(u) - 2F(u) \right), \tag{2.23}$$

and we have the following differential version of Pohozhaev identity (which is very easy to verify for our case of radially symmetric functions)

$$P'(r) = r^{n-1}\left[2nF(u) - (n-2)uf(u) \right] \equiv r^{n-1}I(u), \tag{2.24}$$

where we denote

$$I(u) = 2nF(u) - (n-2)uf(u).$$

To show that $Q(r)$ is positive, it suffices to show that both $P(r)$ and $uf(u) - 2F(u)$ are positive.

Lemma 2.5. *Assume that $uf(u) - 2F(u) > 0$ for $u > b$, (i.e., where $f(u) > 0$), and one of the following three conditions holds*

$$I(u) = 2nF(u) - (n-2)uf(u) > 0 \;\; \textit{for } u > b, \tag{2.25}$$

$$I(u) = 2nF(u) - (n-2)uf(u) < 0 \;\; \textit{for } u > 0, \tag{2.26}$$

$$I(u) < 0 \;\; \textit{for } u \in (0, m), \;\; I(u) > 0 \;\; \textit{for } u > m, \;\; \textit{for some } m > 0. \tag{2.27}$$

Then $P(r) > 0$, for $r \in (0, \eta)$ (i.e., where $f(u) > 0$).

Proof: Since $P(0) = 0$ and $P(1) > 0$, it follows by integrating (2.24) that $P(r) > 0$, for $r \in (0, \eta)$. (In case (2.25) holds, $P(r)$ is increasing on $(0, \eta)$, in case of (2.26), $P(r)$ is decreasing for all r, and in the last case $P(r)$ is positive for all r.) Since $Q(r) > P(r)$ over $(0, \eta)$, the proof follows.

$\diamond$

We now turn to our main example of a cubic.

Lemma 2.6. *Consider positive solutions of the cubic problem*

$$u'' + \frac{n-1}{r}u' + f(u) = 0, \quad u'(0) = 0, \ u(1) = 0, \tag{2.28}$$

where $f(u) = u(u - b)(c - u)$, with constants $c > 2b > 0$. Then any non-trivial solution of the corresponding linearized problem (2.4) is of one sign.

The proof of this lemma will depend on the following lemma of T. Ouyang and J. Shi [145], which is also a good example on how the method of test functions works.

Lemma 2.7. *Define $\alpha = \frac{1}{3}(b + c)$ (i.e., $f''(\alpha) = 0$), and $\rho = \alpha - \frac{f(\alpha)}{f'(\alpha)}$. Then solution of the corresponding linearized problem (2.4), $w(r)$, cannot vanish in the region where $u(r) > \rho$.*

Proof: One begins by observing that

$$f'(u) < \frac{f(u)}{u - \rho}, \quad \text{for all } u \in (\rho, c). \tag{2.29}$$

Indeed, if we consider $g(u) \equiv f(u) - f'(u)(u - \rho)$ on (ρ, c), then $g(\alpha) = 0$ and $g'(u) = -f''(u)(u - \rho)$, so that $g(u)$ has a minimum at α, and (2.29) follows. If we now consider a test function $z(r) = u(r) - \rho$, then $z(r) > 0$, while in view of (2.29),

$$z'' + \frac{n-1}{r}z' + f'(u(r))z = -f(u) + f'(u)(u - \rho) < 0. \tag{2.30}$$

Assuming, on the contrary, that $w(r)$ has roots, let ξ be the first root of $w(r)$ in the region where $u(r) > \rho$. We may assume that $w(0) > 0$, and $w(r) > 0$ on $[0, \xi)$. Combining (2.30) with linearized problem (2.4), we conclude that

$$\left[r^{n-1}(w'z - wz')\right]' > 0 \quad \text{on } [0, \xi).$$

Integrating this over $[0, \xi)$,

$$\xi^{n-1}w'(\xi)z(\xi) > 0,$$

which is a contradiction (since $w'(\xi) < 0$ and $z(\xi) > 0$). $\Diamond$

Proof of Lemma 2.6 We verify that the Theorem 2.3 applies. The function $f(u) = u(u - b)(c - u)$ is convex for $u \in (0, \frac{b+c}{3})$, and concave when $u > \frac{b+c}{3}$. By Lemma 2.7, $w(r)$ cannot vanish in the region where $f(u)$ is concave, so that we only need to verify that $Q(r) > 0$ in the region where $u \in J \equiv (0, \frac{b+c}{3})$ (see Remark 3 to the Theorem 2.3). Compute

$$uf(u) - 2F(u) = u^3 \left[-\frac{1}{2}u + \frac{1}{3}(b + c) \right] > 0 \text{ on } J.$$

Turning to $P(r)$, observe that $P(0) = 0$, $P(1) > 0$, and

$$P'(r) = r^{n-1}I(u) = r^{n-1}u^2 \left[\frac{n-4}{2}u^2 + \frac{-n+6}{3}(b + c)u - 2bc \right]. \quad (2.31)$$

If $n \geq 4$, the quadratic polynomial in the square brackets in (2.31) is negative when $u = 0$, and it changes sign at most once for $u > 0$. We see that $P'(r)$ is either negative, or it changes sign exactly once on $(0, 1)$, with negative values near $r = 1$. In both cases, $P(r)$ is positive. When $n = 3$, we have a quadratic $-\frac{1}{2}u^2 + (b + c)u - 2bc$, which is positive at $u = c$, and since any positive solution of (2.28) satisfies $u(0) < c$, we see that again $P'(r)$ changes sign at most once, and we proceed similarly. In case $n = 2$, we see directly from the definition of $Q(r)$ that it is positive between any two roots of $w(r)$, since $f(u)$ is positive there.

Observing that $K(u) = u \left(\ln f(u) \right)'$, one easily verifies that

$$K'(u) = -\frac{b}{(u - b)^2} - \frac{c}{(c - u)^2} < 0, \quad \text{for all } u \in (0, c) \setminus \{b\}.$$

Since $K(0) = 1$, the condition (2.19) holds.

Finally, by Remark 1 to the Theorem 2.3 and Lemma 2.7, it suffices to verify that $K(\rho) > 1$, to justify the condition (2.17). Since $K(u)$ is decreasing,

$$K(\rho) > K(\alpha) = 1 + \frac{\rho f'(\alpha)}{f(\alpha)} > 1.$$

This completes the proof. $\Diamond$

The following important example is much easier to verify.

Lemma 2.8. *Consider positive solutions of the problem*

$$u'' + \frac{n-1}{r}u' + \lambda(-au + u^p) = 0, \quad u'(0) = 0, \ u(1) = 0, \quad (2.32)$$

with constants $1 < p < \frac{n+2}{n-2}$, and $a > 0$. Then any non-trivial solution of the corresponding linearized problem (2.4) is of one sign.

Proof: Here $uf(u) - 2F(u) > 0$ for all $u > 0$, and one checks that the condition (2.27) of Lemma 2.5 holds. Hence, Lemma 2.5 applies, and $P(r) > 0$. One easily verifies the conditions (2.17) and (2.19), so that the Theorem 2.3 applies, concluding the proof. $\Diamond$

The following example will be used in our discussion of symmetry breaking.

Lemma 2.9. *Consider positive solutions of the problem*

$$u'' + \frac{n-1}{r}u' + \lambda(u^p - a) = 0, \quad u'(0) = 0, \ u(1) = 0, \qquad (2.33)$$

with constants $1 < p < \frac{n+2}{n-2}$ for $n > 2$, and $1 < p < \infty$ for $n = 2$, and a > 0.
Then any non-trivial solution of the corresponding linearized problem (2.4) is of one sign.

Proof: We show that the Theorem 2.3 applies. Indeed, here $K(u) = \frac{pu^p}{u^p-1}$, and $K'(u) = -\frac{ap^2u^{p-1}}{(u^p-a)^2}$. We see that $\lim_{u \to \infty} K(u) = p > 1$, and $K(u)$ is decreasing for all $u > 0$. Since also $K(0) = 0$, we see that the conditions (2.17) and (2.19) hold. Turning to the condition (2.18), compute $uf(u) - 2F(u) = au + \left(1 - \frac{2}{p+1}\right)u^{p+1} > 0$. We also observe that here

$$I(u) = -a(n+2)u + \left[\frac{2n}{p+1} - n + 2\right]u^{p+1}.$$

The quantity in the square bracket is positive, and hence either the second or the third case of the Lemma 2.4 holds. $\Diamond$

Next we consider the case of positive $f(u)$.

Theorem 2.4. *Let $u(r)$ be a positive singular solution of (2.1), and $w(r)$ a non-trivial solution of the linearized problem (2.4). Assume that*

$$f(u) > 0 \ \text{ for all } u > 0. \qquad (2.34)$$

Assume that $w(r)$ cannot vanish in the region where $u(r) \geq \beta$, for some $\beta > 0$ (one condition for this to happen is $f''(u) < 0$ for $u > \beta$).
(i) *Assume that*

$$K(u) \equiv \frac{uf'(u)}{f(u)} \ \text{ is decreasing for } u \in (0, \beta), \text{ and } K(\beta) > 1. \qquad (2.35)$$

Let η denote the point where $u(\eta) = \beta$, and assume also that

$$Q(r) > 0 \ \text{ for } r \in (\eta, 1). \qquad (2.36)$$

Then $w(r)$ cannot have more than one interior root on $(0, 1)$.

(ii) *If, in addition,*

$$K(u) < K(\beta) \;\; for \; u > \beta, \tag{2.37}$$

then $w(r)$ has no interior roots on $(0,1)$, i.e., we may assume that $w(r) > 0$ on $[0,1)$.

Proof: (i) Assume, on the contrary, that $w(r)$ has two interior roots $\eta < \tau < \bar{\tau} < 1$, so that $w(r) > 0$ on $[0,\tau)$ and $w(r) < 0$ on $(\tau, \bar{\tau})$. Again, we consider the function $O(r) = \frac{2}{\gamma'}\xi(r) - \zeta(r)$, with

$$\xi(r) = r^{n-1}\left(u'w - uw'\right),$$

$$\zeta(r) = r^n\left[u'w' + f(u)w\right] + (n-2)r^{n-1}u'w,$$

and a positive constant γ' to be specified. As before, we have

$$O'(r) = \frac{2}{\gamma'}f(u(r))r^{n-1}\left(K(u(r)) - \gamma\right)w(r), \tag{2.38}$$

with $\gamma = \gamma' + 1$. Observe that by our conditions, $K(u(r))$ is an increasing function of r on the interval $(\eta, 1)$, with $K(u(\eta)) > 1$. We now fix γ (and hence γ') by setting $\gamma = K(u(\tau))$. Then, we see from (2.38) that the function $O(r)$ is decreasing on $(0, \bar{\tau})$ (at $r = \tau$, both $w(r)$ and $K(u(r)) - \gamma$ change sign). Since $O(0) = 0$, it follows that

$$O(r) < 0, \;\; for \; r \in (0, \bar{\tau}).$$

Since $\xi(\tau) > 0$, while $\xi(\bar{\tau}) < 0$, there exists a $t \in (\tau, \bar{\tau})$, such that $\xi(t) = 0$. It follows that

$$\zeta(t) = -O(t) > 0. \tag{2.39}$$

On the other hand, since $\frac{u(t)}{w(t)} = \frac{u'(t)}{w'(t)}$, and using (2.36),

$$\zeta(t) = \left[t^n\left(u'w'\frac{u}{w} + f(u)u\right) + (n-2)t^{n-1}u'u\right]\frac{w}{u} = Q(t)\frac{w(t)}{u(t)} < 0,$$

contradicting (2.39).

(ii) We now exclude the possibility that $w(r)$ has exactly one interior root τ, so that $w(r) > 0$ on $[0,\tau)$, and $w(r) < 0$ on $(\tau, 1)$. The argument is similar. Clearly, $\tau \in (\eta, 1)$. Again, we consider the function $O(r)$, and set $\gamma = K(u(\tau)) > 1$. Using the assumption (2.37), we see from (2.38) that $O(r)$ is decreasing on $(0,1)$. (Observe that at $r = \tau$, both $K(u(r)) - \gamma$ and $w(r)$ change sign.) It follows that $O(1) < 0$, i.e., $\zeta(1) = -O(1) > 0$. But $\zeta(1) = u'(1)w'(1) \leq 0$, a contradiction. $\diamond$

Example $f(u) = u^p - u^q$, with $1 < p < q$, and $p < \frac{n+2}{n-2}$ for $n \geq 3$, $1 < p < \infty$ in case $n = 1, 2$. We only need to consider $u \in (0, 1)$, since $f(u) < 0$ for $u > 1$, and so any solution of (2.1) satisfies $0 < u(r) < 1$. One verifies that $K'(u) = -\frac{(p-q)^2 u^{p+q-1}}{(u^p - u^q)^2} < 0$, for all $u \in (0, 1)$. Here $\beta = \left(\frac{p(p-1)}{q(q-1)} \right)^{\frac{1}{q-p}}$, i.e., $f''(u) < 0$ for $u > \beta$. One verifies that $u f(u) > 2F(u)$, where $u(r) < \beta$. (Write $g(u) \equiv u f(u) - 2F(u)$, then $g(0) = g'(0) = 0$ and $g''(u) = u f''(u) > 0$.) We have $2nF(u) - (n-2)u f(u) > \left(\frac{2n}{p+1} - (n-2) \right) f(u) > 0$, for all $u \in (0, 1)$, and then by (2.24), $P(r) > 0$. Then, $Q(r) > P(r) > 0$ on $(\eta, 1)$, verifying the conditions of the theorem.

We shall derive other results on positivity of solutions for the linearized problem in the section on monotone separation of graphs.

2.3   Uniqueness of the solution curve

It turns out that there is only one curve of positive solutions, unless $f(u)$ has multiple positive roots.

Theorem 2.5. *Assume the function $f(u) \in C^2(\bar{R}_+)$ is either positive on $(0, U)$ for some $U \leq \infty$, or it is negative-positive, i.e., $f(u) \leq 0$ on $(0, \beta_1)$, and $f(u) > 0$ on (β_1, U), with $0 < \beta_1 < U \leq \infty$. In case $U < \infty$, assume that $f(U) = 0$ and $f'(U) < 0$. Then all positive solutions of the Dirichlet problem*

$$u''(r) + \frac{n-1}{r} u'(r) + \lambda f(u(r)) = 0 \ \ for \ 0 < r < 1, \ \ u'(0) = u(1) = 0,$$
$$(2.40)$$

with $u(0) \in (0, U)$, lie on a unique solution curve. Moreover, $u(0) \to U$ on one side of the solution curve, as $\lambda \to \infty$.

Proof: The idea is to show that one side of any solution curve has to tend to U, taking up all possible values of $u(0)$ near U, and then by Lemma 2.2 there is no place left for any other solution curve. Let us start at any solution point $(\lambda, u(r))$ of (2.40), and continue it to the side of increasing $u(0)$. By Lemma 2.2, $u(0)$ will be always increasing, and hence it will tend to a limit. Assume first that $U < \infty$. Then by Theorem 2.1, $u(0) \to U$, and we are done. Let now $U = \infty$. If $u(0) \to \infty$, again we are done, so we may assume that $u(0)$ tends to a finite limit, call it L, over a sequence $\{\lambda_m\}$. This sequence $\{\lambda_m\}$ cannot go to infinity, because by Theorem 2.1, the solution $u(r, \lambda)$ would have to tend to a stable root of $f(u)$, and there are no stable roots of $f(u)$. So, the sequence is bounded, and it has a convergent subsequence, for which we keep the same notation, i.e., $\lambda_m \to \bar{\lambda}$. By standard elliptic estimates, $u(r, \lambda_m) \to \bar{u}(r)$, and $(\bar{\lambda}, \bar{u}(r))$ is a solution of (2.40), with $\bar{u}(0) = L$. At the point $(\bar{\lambda}, \bar{u}(r))$, either the Implicit Function Theorem or the Crandall-Rabinowitz Theorem 1.2 applies, giving us a solution curve through this point, with $u(0) > L$ on one side of this curve, contradicting the maximality of L. ◇

2.4 Direction of a turn and exact multiplicity

We are now ready to begin describing the shape of curves of positive solutions for the problem

$$u''(r) + \frac{n-1}{r}u'(r) + \lambda f(u(r)) = 0 \text{ for } 0 < r < 1, \ u'(0) = u(1) = 0,$$
$$(2.41)$$

provided that any non-trivial solution of the corresponding linearized problem

$$w''(r) + \frac{n-1}{r}w'(r) + \lambda f'(u(r))w(r) = 0 \text{ for } 0 \le r < 1, \qquad (2.42)$$
$$w'(0) = w(1) = 0$$

is of one sign. We recall the formula for the direction of the turn, from the Theorem 1.3, which for radial solutions takes the form

$$\tau''(0) = -\lambda \frac{\int_0^1 f''(u(r)) w^3(r) r^{n-1} dr}{\int_0^1 f(u(r)) w(r) r^{n-1} dr}. \qquad (2.43)$$

We shall assume that

$$f(0) \ge 0, \qquad (2.44)$$

which by Hopf's boundary lemma implies that

$$u'(1) < 0. \qquad (2.45)$$

Since we may assume that $w'(1) < 0$, we have by Theorem 1.5

$$\int_0^1 f(u(r)) w(r) r^{n-1} dr = \frac{1}{2\lambda}u'(1)w'(1) > 0.$$

Hence, the numerator in (2.43) determines the sign of the fraction, and the following theorem follows.

Theorem 2.6. *Let $u(r)$ be a singular solution of (2.41), and assume that any non-trivial solution of the linearized problem (2.42) is of one sign, i.e., we may assume that $w(r) > 0$ on $[0, 1)$. If the nonlinearity $f(u)$ is convex (concave), then $\tau''(0) < 0$ (> 0), and the turn to the left (right) occurs at $(\lambda, u(r))$ in the $(\lambda, u(0))$ plane.*

The following results, which were implicit in P. Korman, Y. Li and T. Ouyang [109] (and the earlier ones for the ODE case in [108]), are of central importance.

Theorem 2.7. *Assume that the function $f(u) \in C^2(\bar{R}_+)$ is convex-concave, i.e., for some $\beta > 0$, $f(u)$ is convex on $(0, \beta)$, and concave for $u > \beta$. Assume also that $f(0) \leq 0$. Let $u(r)$ be a singular solution of (2.41), such that $u(0) > \beta$, and moreover (2.45) holds. Assume that any non-trivial solution of the linearized problem (2.42) is of one sign. Then $\tau''(0) > 0$, and the turn to the right occurs at $(\lambda, u(r))$ in the $(\lambda, u(0))$ plane.*

Theorem 2.8. *Assume that the function $f(u) \in C^2(\bar{R}_+)$ is concave-convex, i.e., for some $\beta > 0$, $f(u)$ is concave on $(0, \beta)$, and convex for $u > \beta$. Assume also that $f(0) \geq 0$. Let $u(r)$ be a singular solution of (2.41), such that $u(0) > \beta$, and assume that any non-trivial solution of the linearized problem (2.42) is of one sign. Then $\tau''(0) < 0$, and the turn to the left occurs at $(\lambda, u(r))$ in the $(\lambda, u(0))$ plane.*

The proof of these results involves a lemma and a trick. Here comes the lemma, see [109].

Lemma 2.10. *Assume that $f(0) \leq 0$. If $u(r)$ and $w(r)$ are solutions of (2.41) and (2.42) respectively, then*

$$\int_0^1 f''(u(r)) \, u_r^2 \, w(r) \, r^{n-1} \, dr \leq 0 \,. \tag{2.46}$$

Proof: Consider the function (which was used previously in [182])

$$\theta(r) = r^{n-1} \left(f(u)'w - f(u)w' \right) \,.$$

Here, of course, $f(u)' = f'(u)u'$. Using our equations (2.41) and (2.42), one calculates

$$\theta'(r) = f''(u)u_r^2 \, w(r) \, r^{n-1} \,.$$

Integrating this over $(0, 1)$, and using that $w'(1) < 0$, gives

$$\int_0^1 f''(u(r)) \, u_r^2 \, w(r) \, r^{n-1} \, dr = -f(0)w'(1) \leq 0 \,. \qquad \diamondsuit$$

Proof of the Theorem 2.7   The trick is to compare the integral $\int_0^1 f''(u(r))\, w^3(r)\, r^{n-1}\, dr$ with the one in (2.46).

The functions $-u_r$ and $w(r)$ are both positive on $(0,1)$, and since the first one vanishes at $r = 0$, while $w(1) = 0$ and $-u_r(1) > 0$, they have to intersect on $(0,1)$. We claim that the point of intersection is unique. Differentiating the equation (2.41), we have

$$u_r'' + \frac{n-1}{r} u_r' + \lambda f'(u)u_r = \frac{n-1}{r^2} u_r < 0, \quad \text{for all } r \in (0,1).$$

I.e., u_r is a supersolution of the linearized equation in (2.42), and then $-u_r$ is a subsolution. If $-u_r$ and $w(r)$ intersected more than once, then, considering their values at $r = 0$ and at $r = 1$, we could find an interval $(r_1, r_2) \subset (0,1)$, so that $w(r) < -u_r(r)$ on (r_1, r_2). As $w(r)$ satisfies a linear equation, so do its multiples. We shall now scale w up, to reverse the order between these functions. We can find a constant $\mu > 1$, such that $-u_r \le \mu w$ for all $r \in (r_1, r_2)$, while $-u_r(r_0) = \mu w(r_0)$ and $-u_r'(r_0) = \mu w'(r_0)$ at some $r_0 \in (r_1, r_2)$. We have a subsolution $-u_r$ of the equation (2.42) touching from below a solution μw of the same equation. By the strong maximum principle, this is not possible, proving that $-u_r$ and w intersect exactly once on $(0,1)$.

By scaling $w(r)$, we can make any point on $(0,1)$ to be the unique point of intersection of w and $-u_r$. We show next that by scaling $w(r)$, we can obtain an inequality (using (2.46))

$$\int_0^1 f''(u(r))\, w^3(r)\, r^{n-1}\, dr < \int_0^1 f''(u(r))\, u_r^2\, w(r)\, r^{n-1}\, dr \le 0, \quad (2.47)$$

which, in view of the formula (2.43) for $\tau''(0)$, will complete the proof. Let $\bar{r}$ be such that $u(\bar{r}) = \beta$ (recall that $f(u)$ changes concavity at β). Scale w so that $\bar{r}$ is the unique point of intersection of w and $-u_r$. Then

$$f''(u(r)) < 0 \text{ and } w^2(r) > u_r^2(r) \text{ on } (0, \bar{r})\,;$$

$$f''(u(r)) > 0 \text{ and } w^2(r) < u_r^2(r) \text{ on } (\bar{r}, 1)\,.$$

The inequality (2.47) follows, since we have a pointwise inequality for the corresponding integrands. $\diamond$

The proof of the Theorem 2.8 is similar (the inequalities are reversed in both (2.46) and (2.47)).

We are now ready for exact multiplicity results.

Theorem 2.9. *Assume that the function* $f(u) \in C^2(\bar{R}_+)$ *is convex-concave, and* $f(u)$ *is either positive on* $(0, U)$ *for some* $U \leq \infty$ *and* $f(0) = 0$, *or it is negative-positive, i.e.,* $f(u) \leq 0$ *on* $(0, \beta_1)$, *and* $f(u) > 0$ *on* (β_1, U), *with* $0 < \beta_1 < U \leq \infty$. *In case* $f(0) < 0$, *we assume additionally that* $u'(1) < 0$ *at any singular solution. Assume that any non-trivial solution of the linearized problem (2.42) is of one sign. Then all positive solutions of (2.41), satisfying* $u(0) < U$, *lie on a unique smooth curve, which admits at most one turn to the right in the* $(\lambda, u(0))$ *plane.*

Theorem 2.10. *Assume that the function* $f(u) \in C^2(\bar{R}_+)$ *is concave-convex, and* $f(u)$ *is either positive on* $(0, U)$ *for some* $U \leq \infty$, *or it is negative-positive, i.e.,* $f(u) \leq 0$ *on* $(0, \beta_1)$, *and* $f(u) > 0$ *on* (β_1, U), *with* $0 < \beta_1 < U \leq \infty$ *and* $f(0) = 0$. *Assume that any non-trivial solution of the linearized problem (2.42) is of one sign. Then all positive solutions of (2.41), satisfying* $u(0) < U$, *lie on a unique smooth curve, which admits at most one turn to the left in the* $(\lambda, u(0))$ *plane.*

These results provide us with a "bird's eye view" of the shape of the solution curve. For the exact picture, one also needs to consider the behavior of the solution curve as $u(0) \to 0$, and as $u(0) \to U \leq \infty$, which is relatively easy.

Proof of the Theorem 2.9 By Theorem 2.5, all positive solutions lie on a unique smooth curve. By Theorem 2.7, only turns to the right are possible, and hence at most one such turn may occur. $\diamondsuit$

The Theorem 2.10 is proved similarly.

Our first example is that of a cubic

$$u'' + \frac{n-1}{r}u' + \lambda u(u-b)(c-u) = 0, \quad r \in (0,1), \quad u'(0) = u(1) = 0, \quad (2.48)$$

with constants $0 < b < c$. According to the Theorem 1.9, for the existence of positive solution it is necessary that

$$c > 2b. \quad (2.49)$$

This condition has a geometrical meaning: the area of the positive hump of the cubic (when $b < u < c$) exceeds the area of its negative hump (when $0 < u < b$). Observe that any non-trivial solution of (2.48) is in fact positive, satisfying $0 < u(r) < c$, as follows by the maximum principle. We have the following exhaustive description of the set of positive solutions.

Theorem 2.11. *Assume that the condition (2.49) holds. Then there is a critical* λ_0, *such that for* $\lambda < \lambda_0$ *the problem (2.48) has no positive* *solutions, it has exactly one positive solution at $\lambda = \lambda_0$, and there are exactly two positive solutions for $\lambda > \lambda_0$. Moreover, all solutions lie on a single smooth solution curve, which for $\lambda > \lambda_0$ has two branches, denoted by $u^-(r, \lambda) < u^+(r, \lambda)$, with $u^+(r, \lambda)$ strictly monotone increasing in λ, and $\lim_{\lambda \to \infty} u^+(r, \lambda) = c$ for all $r \in [0, 1)$. For the lower branch, $\lim_{\lambda \to \infty} u^-(r, \lambda) = 0$ for $r \neq 0$.*

Proof: By Lemma 2.6, any solution of the corresponding linearized problem is of one sign, and so according to the Theorem 2.9, only turns to the right are possible, and hence any solution curve has at most one turn. For sufficiently small λ, no positive solution exists, as can be seen by multiplying the PDE version of (2.48) by u, integrating over the unit ball, and using the Poincare's inequality, obtaining

$$\lambda \int_{|x| \leq 1} u^2 (u - b)(c - u) \, dx \geq \lambda_1 \int_{|x| \leq 1} u^2 \, dx \, .$$

Since $(u - b)(c - u) < c_0$, for some $c_0 > 0$ and all u, we have a contradiction for small λ. It is well known that for λ large there is a positive solution, see e.g., R. Gardner and L.A. Peletier [61] (this can also be proved by "shooting", as we explain later on in this book, or by variational methods). We now continue this solution for decreasing λ. The solution curve must turn, as it has no other place to go (solutions are bounded by c; they cannot approach zero by maximum principle, since $f'(u) < 0$ for u small). At the turning point, the Crandall-Rabinowitz Theorem 1.2 applies. Near the turning point we have two strictly ordered branches, since we have positivity for the linearized equation (by the Crandall-Rabinowitz Theorem 1.2, the difference $u^+ - u^-$ is asymptotically proportional to solution of the linearized equation, see Lemma 2.11 below). By the maximum principle, the order is preserved for all λ. By the Theorem 2.1, the upper branch tends to c, and the lower one to zero, for $r \neq 0$. The solution curve is unique, since any other curve would have an upper branch tending to c, contradicting the Lemma 2.2 (since all possible value of $u(0)$ near c are "taken"). Finally, monotonicity of the upper branch follows easily by differentiating the equation (2.48) in λ, and using the maximum principle, see P. Korman, Y. Li and T. Ouyang [109]. $\diamond$

In Figure 2.1, we present a numerically computed bifurcation diagram for the problem (2.48), in case $b = 1$, $c = 5$, $n = 3$. Observe that *Mathematica* chose the point $(3.5, 3.8)$, to originate the axes. Recall that by the Theorem 1.9, we have $F(\alpha) > 0$, where $\alpha = umax = u(0)$, and

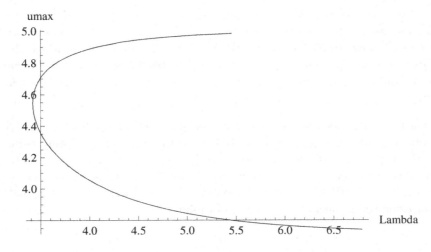

Fig. 2.1   The solution curve for $f(u) = u(u - 1)(5 - u)$, $n = 3$

$F(u) = \int_0^u f(t)\, dt$. Denoting by θ, the positive root of $F(u) = 0$, one computes $\theta \simeq 1.55$. We see that $\lim_{\lambda \to \infty} u(0, \lambda)$ on the lower branch (which is $\simeq 3.7$) is significantly higher than θ. In case $n = 1$, on the lower branch, we have $\lim_{\lambda \to \infty} u(0, \lambda) = \theta$, as discussed in the next chapter.

Our next example is

$$u'' + \frac{n-1}{r} u' + \lambda(-au + u^p) = 0, \quad r \in (0,1), \quad u'(0) = u(1) = 0, \quad (2.50)$$

with constants $a > 0$ and $1 < p < \frac{n+2}{n-2}$. Again, we obtain a complete description of the set of positive solutions.

Theorem 2.12. *For any $\lambda > 0$, the problem (2.50) has a unique positive solution. Moreover, all positive solutions are non-singular, they lie on a single smooth solution curve $u(r, \lambda)$, with $\lim_{\lambda \to 0} u(r, \lambda) = \infty$ for all $r \in [0, 1)$, while $\lim_{\lambda \to \infty} u(r, \lambda) = 0$ for all $r \neq 0$.*

Proof: As before, existence of positive solution for λ large is known (it can be easily proved by "shooting"). We now continue this solution for decreasing λ. By Lemma 2.8, we have positivity for the linearized problem, and then by the Theorem 2.6, only turns to the left are possible at a singular solution (λ_0, u_0), but since the curve has arrived at (λ_0, u_0) from the right, no turns are possible. Similarly, no turns are possible for increasing λ. By Theorem 2.1, $\lim_{r \to \infty} u(r, \lambda) = 0$ for all $r \neq 0$. It follows by Theorem 2.5

that $u(0, \lambda)$ tends to infinity for decreasing λ. By the a priori estimates of B. Gidas and J. Spruck [64], this may happen only if $\lambda \to 0$. $\diamond$

Sometimes solution curve bifurcates either from zero or infinity. The direction of bifurcation is given by the following known proposition, see e.g., Proposition 3.4 in T. Ouyang and J. Shi [145]. One defines $f'(\infty) = \lim_{u \to \infty} \frac{f(u)}{u}$.

Proposition 2.1. **(1)** *Assume that bifurcation of positive solutions from zero occurs for the problem (2.41). If $\frac{f(u)}{u} < f'(0)$ ($\frac{f(u)}{u} > f'(0)$) for small enough $u > 0$, then bifurcation of positive solutions is toward increasing (decreasing) λ.*
(2) *Assume that bifurcation of positive solutions from infinity occurs for the problem (2.41). If $\frac{f(u)}{u} < f'(\infty)$ ($\frac{f(u)}{u} > f'(\infty)$) for large enough $u > 0$, then bifurcation of positive solutions is toward increasing (decreasing) λ.*

Indeed, let us consider bifurcation of positive solutions from infinity, which occurs at $\lambda = \frac{\lambda_1}{f'(\infty)}$. If we have $\frac{f(u)}{u} < f'(\infty)$, this is the same as writing $f(u) = f'(\infty)u - g(u)$, with some $g(u) > 0$. Multiplying (2.41) by ϕ_1, and integrating, we conclude that $\lambda > \frac{\lambda_1}{f'(\infty)}$. (Recall that we denote by λ_1 the first eigenvalue of $-\Delta$ on the unit ball, with zero at the boundary condition, and by $\phi_1 > 0$ the corresponding eigenfunction.)

The next case we consider is

$$u'' + \frac{n-1}{r}u' + \lambda(u + u^p) = 0, \quad r \in (0, 1), \quad u'(0) = u(1) = 0, \quad (2.51)$$

with $1 < p < \frac{n+2}{n-2}$ for $n \geq 6$, and $1 < p < 2$ for $n = 2, 3, 4, 5$.

Theorem 2.13. *For any $0 < \lambda < \lambda_1$ the problem (2.51) has a unique positive solution. Moreover, all positive solutions are non-singular, they lie on a single smooth solution curve $u(r, \lambda)$, which bifurcates from zero at $\lambda = \lambda_1$, and $\lim_{\lambda \to 0} u(0, \lambda) = \infty$.*

Proof: By standard results of M.G. Crandall and P.H. Rabinowitz [39], a curve of positive solutions bifurcates from zero (in the direction of decreasing λ) at $\lambda = \lambda_1$. We now continue the solution curve for decreasing λ, until a possible critical point is reached. It follows from the Theorem 2.18 below (see the Example 1) that any solution of the corresponding linearized problem is positive. Since $f(u) = u + u^p$ is convex, it follows that the solution set near a critical point consists of a curve with a turn to the left. But that is impossible, since the solution curve has arrived at the critical point from the right. It follows that no critical points are encountered, and by the a priori estimates of B. Gidas and J. Spruck [64], we can continue the curve for all $\lambda > 0$.   ◇

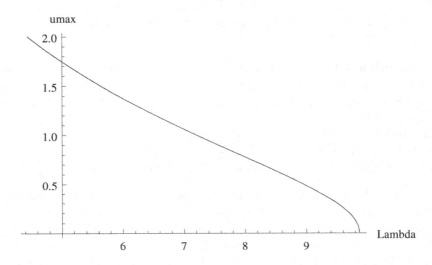

Fig. 2.2   The solution curve for the problem (2.51), with $p = 3$, $n = 3$

In Figure 2.2 we present a computed bifurcation diagram for the problem (2.51). It suggests that the condition, $1 < p < 2$ for $n = 2, 3, 4, 5$, can be probably dropped.

A variation on the theme is the following problem

$$u'' + \frac{n-1}{r}u' + \lambda u + u^p = 0, \quad r \in (0, 1), \quad u'(0) = u(1) = 0, \quad (2.52)$$

with $1 < p < \frac{n+2}{n-2}$ for $n \geq 6$, and $1 < p < 2$ for $n = 2, 3, 4, 5$.

Theorem 2.14. *For any $-\infty < \lambda < \lambda_1$, the problem (2.52) has a unique positive solution. Moreover, all positive solutions are non-singular, they lie on a single smooth solution curve $u(r, \lambda)$, which bifurcates from zero at $\lambda = \lambda_1$, and $\lim_{\lambda \to -\infty} u(0, \lambda) = \infty$.*

Proof: Indeed, bifurcation from zero occurs again at $\lambda = \lambda_1$. We then rescale the problem (2.52) ($u \to \alpha u$, with a suitable α), to put it into the form (2.51) (for this form, the Crandall-Rabinowitz Theorem 1.2 applies), and use the same arguments. At $\lambda = 0$, the problem (2.52) has a positive solution, and for negative λ, we rescale to the problem (2.50), and use Theorem 2.12. $\diamond$

We conclude this section with

$$u'' + \frac{n-1}{r}u' + \lambda\left(u^p - u^q\right) = 0, \quad r \in (0,1), \quad u'(0) = u(1) = 0, \quad (2.53)$$

where $1 < p < q$, and $p < \frac{n+2}{n-2}$ for $n \geq 3$, $1 < p < \infty$ in case $n = 1, 2$. The proof of the following theorem follows the same pattern. We use Theorem 2.4 (see the example afterwards) to prove positivity property for the linearized problem. To prove monotonicity of the branches, we differentiate the equation in λ, and use the strong maximum principle. In Figure 2.3 we present a computational example.

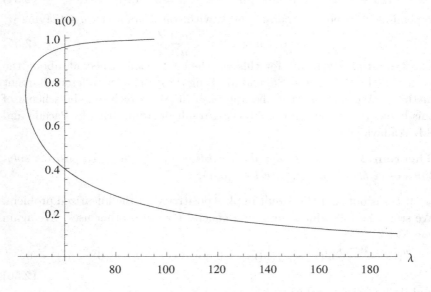

Fig. 2.3   The solution curve for the problem (2.53), with $p = 2$, $q = 3$, and $n = 3$

Theorem 2.15. *There is a critical λ_0, such that for $\lambda < \lambda_0$ the problem (2.53) has no positive solutions, it has exactly one positive solution at $\lambda = \lambda_0$, and there are exactly two positive solutions for $\lambda > \lambda_0$. Moreover, all solutions lie on a single smooth solution curve, which for $\lambda > \lambda_0$ has two branches, denoted by $u^-(r, \lambda) < u^+(r, \lambda)$, with $u^+(r, \lambda)$ strictly monotone increasing in λ, and $\lim_{\lambda \to \infty} u^+(r, \lambda) = 1$ for $r \in [0, 1)$. The lower branch is strictly monotone decreasing in λ, and $\lim_{\lambda \to \infty} u^-(r, \lambda) = 0$ for all r.*

In all of the previous examples we had monotonicity of the stable branch. The last example is a rare case when the unstable (lower) branch is also monotone for all $r \in [0, 1)$. (Our numerical computations show that, say, for a cubic $f(u) = u(u - b)(c - u)$, the lower branch is not monotone in λ.)

2.5 On a class of concave-convex equations

Following the paper by A. Ambrosetti, H. Brezis and G. Cerami [10], there has been a considerable interest in nonlinearities $f(u)$ that are concave for small u, and convex after a certain point, with a prominent example

$$\Delta u + \lambda (u^q + u^p) = 0 \quad \text{for } |x| < 1, u = 0 \quad \text{for } |x| = 1, \tag{2.54}$$

depending on a positive parameter λ, with constants p and q satisfying

$$0 < q < 1 < p < \frac{n+2}{n-2}. \tag{2.55}$$

Exact multiplicity result for this problem was established at about the same time by P. Korman [85] and M. Tang [183], using completely different methods. We shall present the approach in [85], which fits the scheme of this book. It is based on the following result of Adimurthy, F. Pacella and S.L. Yadava [4].

Theorem 2.16. *([4]) Under the condition (2.55), any two positive solutions of (2.54) do not intersect on $[0, 1)$.*

It turns out that this result implies positivity for the linearized problem. We state the following lemma in a general form, for other uses. We again consider

$$u''(r) + \frac{n-1}{r} u'(r) + \lambda f(u(r)) = 0 \quad \text{for } 0 < r < 1, \quad u'(0) = u(1) = 0, \tag{2.56}$$

and its linearized problem

$$w''(r) + \frac{n-1}{r} w'(r) + \lambda f'(u(r)) w(r) = 0 \quad \text{for } 0 \le r < 1, \tag{2.57}$$
$$w'(0) = w(1) = 0.$$

Let us call a *turning point* any singular solution (λ_0, u_0) of (2.56), such that $u'_0(1) < 0$, and near the point (λ_0, u_0) we have two positive solutions to one side of λ_0, and no positive solutions to the other side.

Lemma 2.11. *Assume that any two positive solutions of (2.56) do not intersect on $[0, 1)$. Let (λ_0, u_0) be a turning point of (2.56). Then*

$$w(r) > 0 \quad \text{for all } r \in [0, 1).$$

Proof: The Crandall-Rabinowitz Theorem 1.2 applies at (λ_0, u_0). According to that theorem, near the turning point (λ_0, u_0) the difference of two solutions is asymptotic to a factor of $w(r)$, which implies that $w(r) > 0$ for all $r \in [0, 1)$, if and only if for λ close to λ_0, any two solutions on the solution curve passing through (λ_0, u_0) do not intersect. $\diamond$

We have the following detailed result from P. Korman [85].

Theorem 2.17. *Assume the condition (2.55) is satisfied. Then there is a critical $\lambda_0 > 0$, such that for $\lambda > \lambda_0$ the problem (2.54) has no positive solutions, it has exactly one positive solution for $\lambda = \lambda_0$, and exactly two positive solutions for $\lambda < \lambda_0$. Moreover, all positive solutions lie on a single smooth solution curve, which for $\lambda < \lambda_0$ has two branches denoted by $0 < u^-(r, \lambda) < u^+(r, \lambda)$, with $u^-(r, 0) = 0$, $u^-(r, \lambda)$ strictly monotone increasing in λ for all $r \in [0, 1)$, and $\lim_{\lambda \to 0} u^+(0, \lambda) = \infty$.*

Proof: According to A. Ambrosetti et al [10], there is a $\lambda^* > 0$, such that for $\lambda \in (0, \lambda^*)$ the problem (2.54) has at least two positive solutions. On the other hand, it is easy to see that no positive solutions are possible for large λ. Indeed, we can find a constant $\alpha > 0$ so that $u^q + u^p > \alpha u$ for all $u > 0$. Then from the PDE version of (2.54), denoting by B_1 the unit ball, we have

$$\lambda_1 \int_{B_1} u \phi_1 \, dx = \lambda \int_{B_1} (u^q + u^p) \, \phi_1 \, dx > \lambda \alpha \int_{B_1} u \phi_1 \, dx \,,$$

where $(\lambda_1, \phi_1 > 0)$ is the principal Dirichlet eigenpair of $-\Delta$ on B_1. It follows that no positive solutions exist for $\lambda > \frac{\lambda_1}{\alpha}$. Let now $u(r, \lambda)$ be an arbitrary solution of (2.54), for $\lambda < \lambda^*$. We continue this solution for increasing λ. Since, by the above remarks, this curve cannot be continued indefinitely, it has to reach a turning point (λ_0, u_0), at which the Crandall-Rabinowitz Theorem 1.2 applies. In view of Lemma 2.11, we have $w(r) > 0$ at (λ_0, u_0), where $w(r)$ is a solution of the linearized problem. Observe that the function $u^q + u^p$ is concave for small $u > 0$, and convex for large u, with exactly one point of inflection. It follows by the Theorem 2.10 that only turns to the left are possible in $(\lambda, u(0))$ plane. (Theorem 2.10 followed from the inequality $\int_0^1 f''(u) w^3 r^{n-1} \, dr < \int_0^1 f''(u) w u_r^2 r^{n-1} \, dr$. Here $f(u) = u^q + u^p$. While $f''(u(r))$ has a singularity of order $(1 - r)^{q-2}$ near $r = 1$, by Hopf's boundary lemma $w(r)$ is of order $1 - r$ near $r = 1$, and hence both integrals above converge.)

We now continue both branches for decreasing λ. Since only turns to the left are possible in $(\lambda, u(0))$ plane, both branches continue without any turns. By Lemma 2.2, $u(0, \lambda)$ is increasing on the upper branch and decreasing on the lower one. As $\lambda \downarrow 0$ the lower branch has to tend to zero (there is no other place for it to go, since at $\lambda = 0$ there are no positive solutions). Similarly, on the upper branch $u(0, \lambda)$ has to go to infinity for $\lambda \downarrow \bar{\lambda}$, for some $\bar{\lambda} \geq 0$. By the a priori estimates of B. Gidas and J. Spruck [64], $\bar{\lambda} = 0$.

Since all possible values of $u(0)$ are exhausted, it follows that the solution set of (2.54) consists of one curve, which makes exactly one turn to the left. This implies all of the statements of the theorem, except for monotonicity of the lower branch, which is proved by differentiating (2.54) in λ, and using the strong maximum principle.   ◇

In Figure 2.4, we present the solution curve for (2.54) in case $q = 1/2$, $p = 3$, and $n = 3$, computed using *Mathematica*.

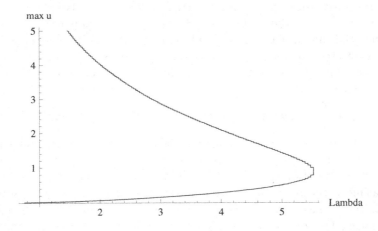

Fig. 2.4   Solution curve for a concave-convex case: $f(u) = u^{1/2} + u^3$, $n = 3$

2.6   Monotone separation of graphs

To treat nonlinearities like $f(u) = u^p + u^q$, with $p > 1$ and $q > 1$, we need to employ a totally different method, developed by L.A. Peletier and J. Serrin [148] and [149], in their study of ground state solutions, and further extended by L. Erbe and M. Tang [53], P. Pucci and J. Serrin [152], and

J. Serrin and M. Tang [167]. We adapt this method to study positivity for the linearized problem, following P. Korman [96].

Since positive solutions of

$$u'' + \frac{n-1}{r}u' + f(u) = 0, \quad r \in (0,1), \quad u'(0) = 0, \ u(1) = 0 \qquad (2.58)$$

are strictly decreasing by B. Gidas, W.-M. Ni and L. Nirenberg [63], we can consider the inverse function $r = r(u)$, which by a Calculus exercise satisfies

$$r''(u) - (n-1)\frac{r'^2(u)}{r(u)} - f(u)r'^3(u) = 0. \qquad (2.59)$$

What is the advantage of this change of variables? The new equation looks more complicated (in fact, it is quasilinear, while the original equation is semilinear). Observe, however, that the new unknown function $r(u)$, and its derivatives, are now "in the open", and not inside of the function f.

If $v(r)$ is another solution of (2.58), then its inverse function $r = s(u)$ satisfies

$$s''(u) - (n-1)\frac{s'^2(u)}{s(u)} - f(u)s'^3(u) = 0. \qquad (2.60)$$

The following fundamental lemma on "monotone separation of graphs" is due to L.A. Peletier and J. Serrin [149].

Lemma 2.12. *([149]) Suppose that $r(u) - s(u) > 0$ on some interval I. Then $r(u) - s(u)$ can have at most one critical point on I. Moreover, this critical point is a strict maximum.*

Proof: If $\bar{u}$ is a critical point of $r(u) - s(u)$, then $r'(\bar{u}) = s'(\bar{u})$, and hence

$$(r - s)''(\bar{u}) = (n-1)\left(\frac{1}{r(\bar{u})} - \frac{1}{s(\bar{u})}\right)r'^2(\bar{u}) < 0.$$

Since any critical point is a point of maximum, it follows that there is at most one critical point. $\diamond$

Assume now there are two intersecting positive solutions of the Dirichlet problem (2.58), $u_1(r)$ and $u_2(r)$. Denote $\alpha_1 = u_1(0)$, and $\alpha_2 = u_2(0)$. By the uniqueness result of L.A. Peletier and J. Serrin [148], we may assume that

$$\alpha_2 > \alpha_1.$$

Let r_0 be the smallest point of intersection, and let $u_0 = u_1(r_0) = u_2(r_0)$. Let $r_1 \leq 1$ be the next point of intersection, with $u_1 = u_1(r_1) = u_2(r_1)$, and $u_1 \geq 0$. Clearly, $r_1(u) - r_2(u) > 0$ on (u_1, u_0). In view of Lemma 2.12, we can find a point $\bar{u} \in (u_1, u_0)$ such that

$$r_1'(\bar{u}) = r_2'(\bar{u}), \quad \text{and} \quad r_1'(u) < r_2'(u) \quad \text{for } u \in (\bar{u}, u_0]. \tag{2.61}$$

It was observed by M. Tang [181] that this inequality continues to hold over the interval (u_0, α_1) too.

Lemma 2.13. *([181]) In the above notation, we have*

$$r_1'(u) < r_2'(u) \quad \text{for } u \in (\bar{u}, \alpha_1). \tag{2.62}$$

Proof: We only need to prove (2.62) over (u_0, α_1). On this interval, the function $r_2(u) - r_1(u) > 0$ can have by Lemma 2.12 at most one critical point. Now, $(r_2 - r_1)(u_0) = 0$ and $(r_2 - r_1)'(u) \to \infty$, as $u \to \alpha_1$. It follows that $r_2 - r_1$ cannot have any critical points at all, i.e., $r_2' - r_1' > 0$ on (u_0, α_1). $\diamondsuit$

The crucial role in the proof of uniqueness is played by an identity of L. Erbe and M. Tang [53], as generalized by P. Pucci and J. Serrin [152]. Defining ([53], [152])

$$P(r) = r^n \left(\frac{1}{2} u'^2(r) + F(u(r)) \right) + n r^{n-1} u'(r) \frac{F(u(r))}{f(u(r))}, \tag{2.63}$$

where as usual, $F(u) = \int_0^u f(t) \, dt$, and $u(r)$ is a solution of (2.58), one shows that ([53], [152])

$$P'(r) = n r^{n-1} u'^2(r) \Phi(u(r)), \tag{2.64}$$

where

$$\Phi(u) \equiv \left(\frac{F(u)}{f(u)} \right)' - \frac{1}{2} + \frac{1}{n}. \tag{2.65}$$

We can also write $P(r) = r^n E(r) + n r^{n-1} u'(r) \frac{F(u(r))}{f(u(r))}$, using the "energy"

$$E(r) = \frac{1}{2} u'^2(r) + F(u(r)).$$

Lemma 2.14. *For any continuous $f(u)$, and any solution $u(r)$ of (2.58)*

$$E(r) > 0 \quad \text{for all } r \in [0, 1).$$

Proof: We have $E'(r) = -\frac{n-1}{r}u'^2(r) < 0$, and $E(1) = \frac{1}{2}u'^2(1) \geq 0$. $\diamond$

As we saw, it is often advantageous to work with $r(u)$ instead of $u(r)$. We shall also be switching back and forth between the two representations. For example, consider the expression $Q(r) \equiv r^{n-1}u'(r)$, which occurs in the definition of $P(r)$. It can be written as

$$Q(u) = \frac{r^{n-1}(u)}{r'(u)}. \qquad (2.66)$$

However, to compute $Q'(u)$, it is easier to observe from (2.58) that $Q'(r) = -r^{n-1}f(u)$, and then by the chain rule

$$Q'(u) = -r^{n-1}(u)f(u)r'(u). \qquad (2.67)$$

Remark One obtains an easy and illuminating proof of the identity (2.64) by writing $P(r) = r^n E(u(r)) + nQ(r)\frac{F(u(r))}{f(u(r))}$, with $E'(r) = -\frac{n-1}{r}u'^2(r)$ and $Q'(r) = -r^{n-1}f(u(r))$.

Given two solutions $r_1(u)$ and $r_2(u)$, (with $\alpha_2 > \alpha_1$), one defines

$$S(u) = \frac{Q_1(u)}{Q_2(u)}, \qquad (2.68)$$

where $Q_i(u)$ denotes $Q(u)$ evaluated at $r_i(u)$, $i = 1, 2$.

Lemma 2.15. *(see* [53], [152], [181]*) Assume that $f(u) > 0$ for $u > \gamma \geq 0$. Then on (γ, α_1) we have $S'(u) < 0$ ($S'(u) > 0$) if and only if $r_1'(u) < r_2'(u)$ ($r_1'(u) > r_2'(u)$).*

Proof: In view of (2.67), we have

$$S'(u) = \left(\frac{r_1}{r_2}\right)^{n-1} \frac{r_2'}{r_1'} f(u)\left(r_2'^2 - r_1'^2\right) < 0,$$

and the proof follows. $\diamond$

In terms of $r(u)$ we rewrite $P(r)$ as

$$P(u) = r^n(u)\left[\frac{1}{2}\frac{1}{r'^2(u)} + F(u)\right] + nQ(u)\frac{F(u)}{f(u)} \qquad (2.69)$$
$$= r^n(u)E(u) + nQ(u)\frac{F(u)}{f(u)},$$

where $Q(u)$ is given by (2.66), and then (2.64) takes the form

$$P'(u) = n\Phi(u)Q(u). \qquad (2.70)$$

Lemma 2.16. *Assume that $\Phi(u) > 0$ for $u > 0$. Then, with $\alpha = u(0)$, we have $P(\alpha) = 0$, and*

$$P(u) > 0 \text{ for } u \in [0, \alpha).$$

Proof: We see directly that $P(\alpha) = 0$, and from (2.70), $P'(u) < 0$ on $(0, \alpha)$. $\diamond$

Recall that in case two positive solutions of (2.58) intersect, we have denoted by (r_0, u_0) the first point of intersection, and there exists a point $\bar{u}$ defined by (2.61). The following theorem is essentially due to L. Erbe and M. Tang [53], see also M. Tang [181], and P. Pucci and J. Serrin [152]. (Some simplification was achieved in P. Korman [96], by introducing the function $\phi(u)$ defined below.)

Theorem 2.18. *(see [53], [181], [152]) Assume that $f(u) > 0$, and also $\Phi(u) > 0$ for $u > 0$. Then any two positive solutions of (2.58) cannot intersect.*

Proof: Given two solutions $r_1(u)$ and $r_2(u)$, one defines (using the functions introduced earlier)

$$\Psi(u) = P_1(u)Q_2(u) - P_2(u)Q_1(u) = r_1^n(u)E_1(u)Q_2(u) - r_2^n(u)E_2(u)Q_1(u),$$

and shows that by two different evaluations, $\Psi(\bar{u})$ is both negative and positive. It follows that the point $\bar{u}$ does not exist, i.e., the solutions do not intersect. Indeed, by Lemma 2.14 and the definition of $\bar{u}$, we have $E_2(\bar{u}) = E_1(\bar{u}) > 0$, while

$$r_1^n(\bar{u})Q_2(\bar{u}) - r_2^n(\bar{u})Q_1(\bar{u}) = r_1^{n-1}(\bar{u})r_2^{n-1}(\bar{u}) \frac{r_1(\bar{u}) - r_2(\bar{u})}{r_1'(\bar{u})} < 0,$$

i.e., $\Psi(\bar{u}) = E_1(\bar{u}) \left[r_1^n(\bar{u})Q_2(\bar{u}) - r_2^n(\bar{u})Q_1(\bar{u}) \right] < 0$.

On the other hand, let us define $\phi(u) = P_1(u)Q_2(\bar{u}) - P_2(u)Q_1(\bar{u})$. Since $P_1(\alpha_1) = 0$ and $P_2(\alpha_1) > 0$ (by Lemma 2.16), we see that $\phi(\alpha_1) = -P_2(\alpha_1)Q_1(\bar{u}) > 0$. By (2.70) and Lemmas 2.13 and 2.15, we see that on the interval $(\bar{u}, \alpha_1)$

$$\phi'(u) = n\Phi(u) \left[Q_1(u)Q_2(\bar{u}) - Q_2(u)Q_1(\bar{u}) \right] \tag{2.71}$$
$$= n\Phi(u)Q_2(u)Q_2(\bar{u}) \left[S(u) - S(\bar{u}) \right] < 0.$$

It follows that $\phi(\bar{u}) > 0$, but $\phi(\bar{u}) = \Psi(\bar{u}) < 0$, a contradiction. $\diamond$

The condition $\Phi(u) > 0$ may be also written as

$$\frac{f'(u)F(u)}{f^2(u)} < \frac{n+2}{2n}. \tag{2.72}$$

It holds for many positive functions $f(u)$.

Example 1 $f(u) = u^p + u^q$, with $1 \leq p < q$, and $q < \frac{n+2}{n-2}$, for $n > 2$. We show that (2.72) holds, provided that $q - p < 1$. Compute

$$f'(u)F(u) = \frac{p}{p+1}u^{2p} + \left(\frac{q}{p+1} + \frac{p}{q+1}\right)u^{p+q} + \frac{q}{q+1}u^{2q}.$$

Clearly, $\frac{p}{p+1} < \frac{q}{q+1}$, and we also have

$$\frac{q}{p+1} + \frac{p}{q+1} < 2\frac{q}{q+1},$$

since $(q-p)^2 < q - p$ by our assumptions. It follows that $f'(u)F(u) < \frac{q}{q+1}f^2(u)$, i.e.,

$$\frac{f'(u)F(u)}{f^2(u)} < \frac{q}{q+1} < \frac{n+2}{2n},$$

since $q < \frac{n+2}{n-2}$. Of course, for $n \geq 6$, the condition $q - p < 1$ holds automatically.

Example 2 $f(u) = u^p + a$, with $p > 1$. Calculate

$$f'(u)F(u) - \frac{n+2}{2n}f^2(u) = \left(\frac{p}{p+1} - \frac{n+2}{2n}\right)u^{2p} + a\left(p - \frac{n+2}{n}\right)u^p - \frac{n+2}{2n}a^2.$$

On the right hand side we have a quadratic polynomial in u^p, whose leading coefficient is negative, provided that $p < \frac{n+2}{n-2}$. This quadratic is negative for all u, and any $a > 0$, if

$$\left(p - \frac{n+2}{n}\right)^2 < -4\left(\frac{p}{p+1} - \frac{n+2}{2n}\right)\frac{n+2}{2n}. \tag{2.73}$$

The difference between the right hand side and the left hand side of (2.73) is $\frac{p^2(4+n-np)}{n(p+1)}$. It follows that (2.73) is satisfied, and hence the condition (2.72) holds for

$$1 < p < \frac{n+4}{n}.$$

(Observe that $\frac{n+4}{n} < \frac{n+2}{n-2}$, for all $n \geq 3$.)

Example 3 $f(u) = (u + \gamma)^p$, with constants $\gamma > 0$, and $1 < p < \frac{n+2}{n-2}$. The condition (2.72) is easy to verify.

We have given some conditions under which any two positive solutions of (2.58) cannot intersect. Recall that by Lemma 2.11 this implies that any non-trivial solution of the corresponding linearized problem is of one sign, i.e., it may be assumed to be positive. Once we have positivity for the linearized problem, our methods imply the following exact multiplicity results (see e.g., the proof of Theorem 2.12).

Theorem 2.19. *For any $\lambda > 0$, the problem*

$$u'' + \frac{n-1}{r}u' + \lambda(u^p + u^q) = 0, \quad r \in (0,1), \quad u'(0) = 0, \ u(1) = 0, \quad (2.74)$$

with $1 < p, q < \frac{n+2}{n-2}$, has a unique positive solution. Moreover, all solutions lie on a single smooth solution curve $u(r, \lambda)$, which tends to infinity as $\lambda \to 0$, and to zero when $\lambda \to \infty$, for all $r \in [0,1)$.

Uniqueness part was proved previously by L. Erbe and M. Tang [53], but by a considerably more complicated method. Under a more restrictive condition $1 < p, q < \frac{n}{n-2}$, this result was proved in T. Ouyang and J. Shi [145].

Theorem 2.20. *Consider the problem*

$$u'' + \frac{n-1}{r}u' + \lambda(u + \gamma)^p = 0, \quad r \in (0,1), \quad u'(0) = 0, \ u(1) = 0,$$

with constants $\gamma > 0$, and $1 < p < \frac{n+2}{n-2}$. Then all positive solutions lie on a single smooth solution curve $u(r, \lambda)$, which begins at $(\lambda = 0, u = 0)$, it makes exactly one turn to the left at some $\lambda_0 > 0$, and then tends to infinity as $\lambda \to 0$. Correspondingly, we have exactly two positive solutions for $\lambda \in (0, \lambda_0)$, exactly one solution at $\lambda = \lambda_0$, and no solutions for $\lambda > \lambda_0$.

Under a more restrictive condition $1 < p < \frac{n}{n-2}$, this result was proved in T. Ouyang and J. Shi [145].

We conclude this section with another application of the method of monotone separation of graphs. Recall that we have denoted $E(r) = \frac{1}{2}u'^2(r) + F(u(r))$, and that $E'(r) = -\frac{n-1}{r}u'^2(r)$. Compute

$$\left(r^k E(r)\right)' = r^{k-1}\left[\frac{k}{2}u'^2 + kF(u)\right] - (n-1)r^{k-1}u'^2.$$

If we choose here $k = 2(n-1)$, we obtain the following identity of L.A. Peletier and J. Serrin [149]

$$\left(r^{2(n-1)}E(r)\right)' = 2(n-1)r^{2n-3}F(u(r)). \quad (2.75)$$

We shall need the following result, which is essentially due to L.A. Peletier and J. Serrin [148], who considered ground state solutions. However, their argument works equally well for Dirichlet problems, as was observed by M.K. Kwong and L. Zhang [122], and the details were later written up in P. Korman [96].

Theorem 2.21. *([148]) Assume that $F(u) < 0$ for $0 < u < \theta$. Let r_0 be a point of intersection of two positive solutions of (2.58), $u_1(r)$ and $u_2(r)$, with $u_0 = u_1(r_0) = u_2(r_0)$. Then*

$$u_0 > \theta. \tag{2.76}$$

Proof: Assume that $u_1'(r_0) > u_2'(r_0)$. Recall that given the intersection point (r_0, u_0), we can find $\bar{u}$, satisfying (2.61). Let $\bar{r}_1$ and $\bar{r}_2$ be defined by $u_1(\bar{r}_1) = u_2(\bar{r}_2) = \bar{u}$. We now integrate the identity (2.75), written for the first solution, over the interval $(r_0, \bar{r}_1)$ (denoting $E_1(r) = \frac{1}{2}{u_1'}^2(r) + F(u_1(r))$), and switch to the u variable, $u = u_1(r)$, with $dr = r_1'(u)du$,

$$\bar{r}_1^{2(n-1)} E_1(\bar{r}_1) - r_0^{2(n-1)} E_1(r_0) = -2(n-1) \int_{\bar{u}}^{u_0} r_1^{2n-3}(u) F(u) r_1'(u)\, du.$$

Similarly, we integrate the identity (2.75), written for the second solution, over the interval $(r_0, \bar{r}_2)$, and then subtract from the first formula the second one (with $E_2(r) = E(u_2(r))$, and $k = 2(n-1)$). Observing that $E_1(\bar{r}_1) = E_2(\bar{r}_2)$), we have

$$\left(\bar{r}_1^k - \bar{r}_2^k\right) E_1(\bar{r}_1) + r_0^k \left(E_2(r_0) - E_1(r_0)\right) \tag{2.77}$$
$$= k \int_{\bar{u}}^{u_0} \left(r_2^{k-1} r_2'(u) - r_1^{k-1} r_1'(u)\right) F(u)\, du.$$

The first term on the left is positive by Lemma 2.14, and the second one is positive, since $u_2(r)$ is steeper than $u_1(r)$ at r_0. By (2.61), we have

$$r_1^{k-1} r_1'(u) < r_2^{k-1} r_2'(u) < 0, \text{ for } u \in (\bar{u}, u_0).$$

Hence, if we assume that $u_0 \leq \theta$, then $F(u) < 0$, and then the right hand side of (2.77) is negative, giving a contradiction. $\diamond$

Corollary If, moreover $F(\theta) = 0$, and $f(u) > 0$, $\Phi(u) > 0$ for $u > \theta$, then any two positive solutions of (2.58) cannot intersect more than once.

Combining the last result with Lemmas 2.7 and 2.11, we get the following general condition for positivity of solution of the linearized problem.

Theorem 2.22. *Assume that $F(u) < 0$ for $0 < u < \theta$, while*

$$f'(u) < \frac{f(u)}{u - \theta}, \text{ for } u > \theta.$$

If (λ, u) is a turning point of (2.58), then solution of the corresponding linearized problem may be assumed to be positive on $[0, 1)$.

2.7   The case of polynomial $f(u)$ in two dimensions

The results of the preceeding section depended on a particular form of the nonlinearity $f(u)$, which was a sum of two powers. In case of the space dimension $n = 2$, we may allow $f(u)$ to be a sum of arbitrarily many powers, by using a completely different method, see P. Korman [83]. We consider positive solutions on a unit ball in R^2, which satisfy

$$u'' + \frac{1}{r}u' + f(u) = 0 \quad \text{for } 0 < r < 1, \quad u'(0) = u(1) = 0. \tag{2.78}$$

Corresponding linearized problem is

$$w'' + \frac{1}{r}w' + f'(u)w = 0 \quad \text{for } 0 < r < 1, \quad w'(0) = w(1) = 0. \tag{2.79}$$

We consider a class of polynomial nonlinearities

$$f(u) = \sum_{i=1}^{k} a_i u^{p_i} + \alpha,$$

with constants $a_i > 0$, $1 < p_1 < p_2 < \cdots < p_k$, and $\alpha \geq 0$. Following W.-M. Ni and R. Nussbaum [138], we proved the following result in P. Korman [83].

Theorem 2.23. (i) *Assume that $\alpha > 0$, and $1 < p_k \leq 4$. Then any non-trivial solution of the linearized problem (2.79) is of one sign.*
(ii) *Assume that $\alpha = 0$, and $p_k \leq p_1 + 4$. Then the linearized problem (2.79) has only the trivial solution.*

This led us to the following exact multiplicity result. The uniqueness in its second part is due originally to W.-M. Ni and R. Nussbaum [138].

Theorem 2.24. *Consider the problem*

$$\Delta u + \lambda f(u) = 0 \quad \text{for } |x| < 1, \quad u = 0 \text{ for } |x| = 1, \tag{2.80}$$

depending on a positive parameter λ, in case $n = 2$. In the conditions of the first part of the Theorem 2.23, there is a critical $\lambda_0 > 0$, such that the problem (2.80) has exactly two positive solutions for $0 < \lambda < \lambda_0$, it has exactly one positive solution at $\lambda = \lambda_0$, and no solutions for $\lambda > \lambda_0$. Moreover, all positive solutions lie on a single smooth solution curve. This curve begins at $(\lambda = 0, u = 0)$, it makes a single turn to the left at $\lambda = \lambda_0$, and then it tends to infinity, as $\lambda \to 0$.

In the conditions of the second part of the Theorem 2.23, the problem (2.80) has exactly one positive solution for any λ. Moreover, all positive solutions are non-degenerate, and lie on a single smooth solution curve, with $u(0, \lambda) \to 0$ $(u(0, \lambda) \to \infty)$, as $\lambda \to \infty$ $(\lambda \to 0)$.

Proof: The proof of the first part follows the familiar scheme, so we just sketch it. When $\lambda = 0$, there is a trivial solution $u = 0$, which we continue for small λ, by using the Implicit Function Theorem. This solution curve cannot be continued indefinitely for increasing λ, since no solution exists for large λ, as follows by the formula (1.17) of the preceding chapter. Observe that the solution curve cannot go to infinity at a positive λ, as follows from the proof of the Theorem 1.4 in J. Hempel [69]. It follows that the solution curve must reach a turning point (λ_0, u_0). Since $f(u)$ is convex, and solutions of the linearized problem are positive, a turn to the left must occur at any turning point. This means that after the initial turn at (λ_0, u_0), no more turns may occur, and so the solution curve always travels to the left. By the above mentioned estimates of J. Hempel [69], solution curve cannot go to infinity at a positive λ, hence it has to go infinity as $\lambda \to 0$, since at $\lambda = 0$ there are no non-trivial solutions.

Turning to the second part, it is easy to see that solutions on any curve go zero as $\lambda \to \infty$, and they go to infinity as $\lambda \to 0$. It follows that one solution curve "takes up" all possible values of $u(0)$. We conclude that there is exactly one solution curve (existence of solution is well known, see W.-M. Ni and R. Nussbaum [138]), and since the solution curve admits no turns, we conclude the uniqueness of solution. $\diamond$

Remark.   W.-M. Ni and R. Nussbaum [138] had conjectured that the condition $p_k \leq p_1 + 4$ in the second part of the Theorem 2.24 is not necessary. As far as we know, this is still unknown. Similarly, it seems likely that condition $p_k \leq 4$ in the first part of the Theorem 2.24 can be dropped.

2.8   The case when $f(0) < 0$

So far, we have considered positive solutions of the Dirichlet problem
$$u'' + \frac{n-1}{r}u' + \lambda f(u) = 0, \quad r \in (0,1), \quad u'(0) = 0, \ u(1) = 0 \qquad (2.81)$$
for nonlinearities satisfying $f(0) \geq 0$. This condition, and the Hopf's boundary lemma imply that $u'(1) < 0$. Combining this with fact that positive solutions are decreasing, we concluded that solutions $u(r, \lambda)$ remain positive for all λ. Things are different when
$$f(0) < 0. \qquad (2.82)$$
Positive solutions $u(r, \lambda)$ may develop zero slope at the boundary point $r = 1$, at some $\bar{\lambda}$, after which radial sign-changing solutions and non-radial (symmetry breaking) solutions appear. In this section we shall follow

positive solutions up to the point $\bar{\lambda}$, and we shall discuss symmetry breaking in the next section.

We consider again the linearized problem

$$w''(r) + \frac{n-1}{r}w'(r) + \lambda f'(u(r))w(r) = 0 \text{ for } 0 \le r < 1, \qquad (2.83)$$
$$w'(0) = w(1) = 0 .$$

Again, positivity of $w(r)$ will play an important role, used not just to compute the direction of bifurcation, but also to prove that solution curve arrives at the point where slope is zero at $r = 1$, while traveling forward in λ. We need the following two lemmas.

Lemma 2.17. *Assume that $u_r(1, \bar{\lambda}) = 0$. Assume that any non-trivial solution $w(r)$ of the linearized problem (2.83) at $\lambda = \bar{\lambda}$ has to be positive (of one sign). Then solution $u(r, \bar{\lambda})$ is radially non-singular, i.e., the linearized problem (2.83) has no nontrivial solutions at $(\bar{\lambda}, u(r, \bar{\lambda}))$.*

Proof: Assume, on the contrary, that (2.83) has a nontrivial solution, $w(r) > 0$. Differentiate the equation (2.81)

$$u_r'' + \frac{n-1}{r}u_r' + \lambda f'(u)u_r = \frac{n-1}{r^2}u'. \qquad (2.84)$$

From the equations (2.83) and (2.84), we obtain

$$\left[r^{n-1}\left(u''w - u'w'\right) \right] \big|_0^1 = \int_0^1 (n-1)r^{n-3}u'w\, dr.$$

The integral on the right is negative, while the quantity on the left is equal to zero, a contradiction. $\diamond$

Differentiating the equation (2.81) in λ, we see that the function $u_\lambda(r, \lambda)$ satisfies

$$u_\lambda'' + \frac{n-1}{r}u_\lambda' + \lambda f'(u)u_\lambda = -f(u). \qquad (2.85)$$

It turns out that a point where positivity is lost, we can write down explicitly non-trivial solutions of the linearized problem (2.83).

Lemma 2.18. *Assume that $u_r(1, \bar{\lambda}) = 0$. Then the linearized problem (2.83) at $u(r, \bar{\lambda})$ has a non-trivial solution*

$$w(r) = ru_r(r, \bar{\lambda}) - 2\bar{\lambda}u_\lambda(r, \bar{\lambda}) . \qquad (2.86)$$

Proof: Follows by a direct computation. ◇

Definition We say that a positive solution branch $u(r, \lambda)$ *loses positivity forward* at some $\bar{\lambda}$, if $u_r(1, \bar{\lambda}) = 0$, and the solution branch arrives at the point $(\bar{\lambda}, u(r, \bar{\lambda}))$ with λ increasing (i.e., for $\lambda < \bar{\lambda}$ solutions are positive with $u_r(1, \lambda) < 0$).

Theorem 2.25. *Assume that any non-trivial solution of the linearized problem (2.83) is of one sign on $(0, 1)$. Assume that $u(0, \lambda)$ is decreasing on a branch, as we arrive at the point where $u_r(1, \bar{\lambda}) = 0$. Then positivity can be lost only forward at $\bar{\lambda}$.*

Proof: Assume, on the contrary, that positivity is lost backward, i.e., $u(r, \lambda)$ arrives at $\bar{\lambda}$ when λ is decreasing. We claim that

$$u_\lambda(r, \bar{\lambda}) \text{ is positive near } r = 1. \tag{2.87}$$

Indeed, by the definition of $\bar{\lambda}$, it is clear that $u_\lambda(r, \bar{\lambda})$ cannot be negative on an interval containing $r = 1$. So, if $u_\lambda(r, \bar{\lambda})$ failed to be positive on an interval containing $r = 1$, it would have to have infinitely many sign changes near $r = 1$. Let $\mu_n \to 1$ be points of positive local maximums of $u_\lambda(r, \bar{\lambda})$, i.e., $u_\lambda(\mu_n, \bar{\lambda}) > 0$ and $u'_\lambda(\mu_n, \bar{\lambda}) = 0$. We now evaluate the equation (2.85) at $r = \mu_n$. The first term on the left is non-positive, the second one is zero, while the third one tends to zero. The right hand side tends to $-f(0) > 0$, a contradiction, proving the claim.

By Lemma 2.18, $w(r) = ru_r(r, \bar{\lambda}) - 2\bar{\lambda}u_\lambda(r, \bar{\lambda})$. By Lemma 2.17, $w(r) \equiv 0$, i.e.,

$$u_\lambda(r, \bar{\lambda}) = \frac{1}{2\bar{\lambda}} ru_r(r, \bar{\lambda}) < 0, \text{ for all } r, \tag{2.88}$$

contradicting (2.87). ◇

Theorem 2.25 provides us with an understanding of the lower branch of any solution curve. The formula (2.88) gives the tangential direction for radial solutions, at the point where positivity is lost. We can now obtain various uniqueness and exact multiplicity results, by following the scheme outlined in the previous sections. In particular, we obtain a complete understanding of bifurcation diagrams for concave $f(u)$.

Consider the case

$$f''(u) < 0 \text{ for } u > 0, \text{ and } f(0) < 0. \tag{2.89}$$

We may assume that $f(u)$ has a root $u_1 > 0$, since if $f(u)$ is negative, the problem (2.81) has no positive solutions. Consistent with (2.89), there are two possibilities: $f(u)$ has a second root, or $f(u) > 0$ for $u > u_1$. The last case can be divided into two sub-cases: $\lim_{u \to \infty} \frac{f(u)}{u} = 0$, or $f(u)$ is asymptotically linear as $u \to \infty$. The following results cover all of these possibilities. Related results were established by other methods in A. Castro, S. Gadam and R. Shivaji [29], see also T. Ouyang and J. Shi [145], and P. Korman [82].

Theorem 2.26. *Assume that the condition (2.89) holds, and $f(u_1) = f(u_2) = 0$ for some $0 < u_1 < u_2$. If the solution set of the problem (2.81) is non-empty, then all solutions lie on a unique solution curve. Moreover, there are two critical numbers $\lambda_0 < \bar{\lambda}$, such that for $\lambda < \lambda_0$ the problem (2.81) has no positive solutions, it has exactly one positive solution for $\lambda = \lambda_0$, and for $\lambda \geq \bar{\lambda}$, and there are exactly two positive solutions for $\lambda_0 < \lambda < \bar{\lambda}$. Moreover, all positive solutions lie on a single smooth solution curve, which for $\lambda > \lambda_0$ has two branches denoted by $0 < u^-(r, \lambda) < u^+(r, \lambda)$, with $u^+(r, \lambda)$ strictly monotone increasing in λ, and $\lim_{\lambda \to \infty} u^+(r, \lambda) = u_2$, for all $r \in [0, 1)$. For the lower branch, $u^-(0, \lambda)$ is strictly monotone decreasing in λ, and $u_r^-(1, \bar{\lambda}) = 0$. For $\lambda > \bar{\lambda}$, the lower branch continues as a curve of radial sign changing solutions, negative near $r = 1$.*

Proof: As we have seen before (see Lemma 2.7), under the condition (2.89), any solution of the corresponding linearized problem is of one sign. This implies that only one turn to the right is possible on the solution curve, and that positivity of solution can be lost only forward on the lower branch, by the Theorem 2.25. As before, we obtain two solution branches emanating from λ_0. Solutions on the upper branch are increasing in λ (using the maximum principle, it is easy to show that $u_\lambda(r, \lambda) > 0$ for all $r \in [0, 1)$, see [82]), and hence they stay positive. Solutions on the lower branch have no place to go for large λ, since $f(u)$ has no stable roots besides u_2. Hence, at some $\bar{\lambda}$ solutions on the lower branch develop a zero slope at $r = 1$. Since $u^-(r, \bar{\lambda})$ is radially non-singular by Lemma 2.17, we can continue radially symmetric solution for $\lambda > \bar{\lambda}$, and by (2.88), solutions become sign changing. $\diamond$

Similarly, we have the following result.

Theorem 2.27. *Assume that the condition (2.89) holds, $f(u_1) > 0$ and $f(u) > 0$ for $u > u_1$, and $\lim_{u \to \infty} \frac{f(u)}{u} = 0$. Then the problem (2.81) has a unique curve of positive solutions. Moreover, there are two critical numbers $\lambda_0 < \bar{\lambda}$, such that for $\lambda < \lambda_0$ the problem (2.81) has no positive* *solutions, it has exactly one positive solution for $\lambda = \lambda_0$, and for $\lambda \geq \bar{\lambda}$, and there are exactly two positive solutions for $\lambda_0 < \lambda < \bar{\lambda}$. Moreover, all positive solutions lie on a single smooth solution curve, which for $\lambda > \lambda_0$ has two branches denoted by $0 < u^-(r, \lambda) < u^+(r, \lambda)$, with $u^+(r, \lambda)$ strictly monotone increasing in λ, and $\lim_{\lambda \to \infty} u^+(r, \lambda) = \infty$. For the lower branch, $u^-(0, \lambda)$ is strictly monotone decreasing in λ, and $u_r^-(1, \bar{\lambda}) = 0$. For $\lambda > \bar{\lambda}$, the lower branch continues as a curve of radial sign changing solutions, negative near $r = 1$.*

We have the following result for the asymptotically linear case.

Theorem 2.28. *Assume the condition (2.89) holds, $f(u_1) = 0$, and for some constant $a > 0$*

$$\lim_{u \to \infty} \frac{f(u)}{u} = a \,.$$

Then the problem (2.81) has a curve of positive solutions, bifurcating from infinity at $\lambda = \frac{\lambda_1}{a}$, which travels without any turns for increasing λ, developing zero slope at the boundary and losing positivity at some $\lambda = \bar{\lambda}$. Consequently, the problem (2.81) has a unique positive solution for $\frac{\lambda_1}{a} < \lambda < \bar{\lambda}$, and no positive solutions for other $\lambda > 0$. (See Figure 2.5 for an example.)

Proof: By standard results, bifurcation from infinity happens at $\lambda = \bar{\lambda}$, forward in λ, see Proposition 2.1. Since only turns to the right are possible on the solution curve, it follows that no turns occur. The rest of the proof is the same as for the preceding theorems. $\diamond$

Remark Under the condition $f(0) < 0$, it was conceivable that the problem

$$\Delta u + \lambda f(u) = 0 \quad \text{for } x \in B, \quad u = 0 \text{ on } \partial B$$

on a unit ball $B \subset R^n$ could have non-negative solutions (in case $f(0) \geq 0$, Hopf's boundary lemma rules out such a possibility). However, A. Castro and R. Shivaji [30] proved that any non-negative solution is in fact positive, and hence radially symmetric by [63].

2.9 Symmetry breaking

We assume throughout this section that

$$f(0) < 0 \,, \tag{2.90}$$

which is a necessary condition for bifurcation of symmetry breaking solutions, off the curve of positive ones. Indeed, if $f(0) \geq 0$, then the slope of

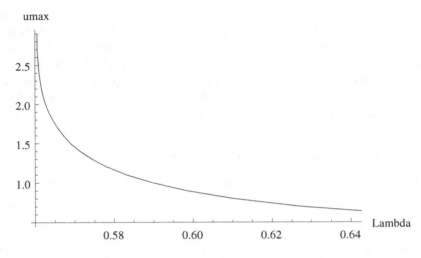

Fig. 2.5   Solution curve for (2.81), $f(u) = 1 + 10u - 2e^{-u}$, $n = 2$

solutions at the boundary is negative, in view of Hopf's boundary lemma. Since radial solutions are strictly decreasing, positive solutions stay positive for all λ, and hence radial by [63]. We consider the original PDE problem

$$\Delta u + \lambda f(u) = 0 \quad \text{for } |x| < 1, \quad u = 0 \text{ on } |x| = 1, \qquad (2.91)$$

and study both radially symmetric and non-radial solutions. As we just saw, usually one branch of solution curve loses positivity at some $\lambda = \bar{\lambda}$, and then continues as a curve of radial sign changing solutions. We call this branch $U(r, \lambda)$. We then set $v(x, \lambda) = u(x, \lambda) - U(r, \lambda)$, and show that in the resulting equation for $v(x, \lambda)$, "bifurcation from zero" of non-radial solutions occurs at $\lambda = \bar{\lambda}$. The following two lemmas provide a condition for the crucial "transversality" condition of the Crandall-Rabinowitz [39] theorem to hold at the points where symmetry breaking is possible. The proofs are taken from P. Korman [84]. They were motivated by M. Ramaswamy [155], who considered the one-dimensional case. Recall that we denote derivatives in r alternatively as u' or u_r, so that u'_λ denotes the mixed partial $u_{\lambda r}$.

Lemma 2.19. *Assume that a branch of radial solutions $u(r, \lambda)$ of*

$$u'' + \frac{n-1}{r}u' + \lambda f(u) = 0, \quad r \in (0, 1), \quad u'(0) = 0, \; u(1) = 0 \qquad (2.92)$$

is defined for $\lambda < \bar{\lambda}$, or for $\lambda > \bar{\lambda}$. Assume that $u_r(1, \bar{\lambda}) = 0$. Then

$$\bar{\lambda} \int_0^1 f''(u) u_\lambda u'^2 r^{n-1} \, dr + \int_0^1 f'(u) u'^2 r^{n-1} \, dr = -\bar{\lambda} f(0) u'_\lambda(1, \bar{\lambda}). \quad (2.93)$$

Proof: Let us assume first that $n > 2$, as there is a slight complication in case $n = 2$. Denoting the left side of (2.93) by $\bar{\lambda} I + J$, we express (integrating by parts, and using that $u_\lambda(1, \lambda) = 0$)

$$I \equiv \int_0^1 f''(u) u_\lambda u'^2 r^{n-1} \, dr = \int_0^1 f'(u)' u_\lambda u' r^{n-1} \, dr \quad (2.94)$$
$$= - \int_0^1 \left[f'(u) u'' u_\lambda r^{n-1} + f'(u) u' u'_\lambda r^{n-1} + f'(u) u' u_\lambda (n-1) r^{n-2} \right] dr$$
$$\equiv I_1 + I_2 + I_3.$$

Proceeding similarly, we express

$$I_2 \equiv - \int_0^1 f(u)' u'_\lambda r^{n-1} \, dr = \int_0^1 f(u) u''_\lambda r^{n-1} \, dr + (n-1) \int_0^1 f(u) u'_\lambda r^{n-2} \, dr$$
$$- f(0) u'_\lambda(1) \, ;$$

$$I_3 = \int_0^1 \left[(n-1) f(u) u'_\lambda r^{n-1} \, dr + (n-1)(n-2) f(u) u_\lambda r^{n-2} \right] dr \, .$$

Using these expressions in (2.94), we have

$$I = \int_0^1 [-f'(u) u'' u_\lambda r^{n-1} + f(u) u''_\lambda r^{n-1} + 2 f(u) u'_\lambda (n-1) r^{n-2} \quad (2.95)$$
$$+ f(u) u_\lambda (n-1)(n-2) r^{n-3}] \, dr - f(0) u'_\lambda(1).$$

From the corresponding equations, we express

$$r^{n-1} u'' = -(n-1) r^{n-2} u' - \bar{\lambda} f r^{n-1},$$

$$r^{n-1} u''_\lambda = -(n-1) r^{n-2} u'_\lambda - \bar{\lambda} f' u_\lambda r^{n-1} - f r^{n-1}.$$

Using these relations, we rewrite the first two terms in (2.95) as follows

$$\int_0^1 \left[-f'(u) u'' u_\lambda r^{n-1} + f(u) u''_\lambda r^{n-1} \right] dr \quad (2.96)$$
$$= \int_0^1 \left[f'(u) u' u_\lambda (n-1) r^{n-2} - f(u) u'_\lambda (n-1) r^{n-2} - f^2(u) r^{n-1} \right] dr =$$
$$\int_0^1 \left[-2 f(u) u'_\lambda (n-1) r^{n-2} - f(u) u_\lambda (n-1)(n-2) r^{n-3} - f^2(u) r^{n-1} \right] dr.$$

Using this in (2.95), we have

$$I = - \int_0^1 f^2(u) r^{n-1} \, dr - f(0) u'_\lambda(1) \, . \quad (2.97)$$

Turning to the second integral in (2.93), we express (by using that $u'(1, \bar{\lambda}) = 0$ on the first step, and the equation (2.92) on the last step)

$$J \equiv \int_0^1 f'(u)u'^2 r^{n-1} \, dr = \int_0^1 f(u)' u' r^{n-1} \, dr \tag{2.98}$$
$$= -\int_0^1 \left[f(u)u'' r^{n-1} + (n-1)f(u)u' r^{n-2} \right] \, dr = \bar{\lambda} \int_0^1 f^2(u) r^{n-1} \, dr.$$

Putting together (2.97) and (2.98), we conclude the proof of the lemma in case $n > 2$.

Turning to the case $n = 2$, we proceed similarly. There is now an extra term $-f(u(0))u_\lambda(0)$ in the expression for I_3, and consequently in (2.95). However, in (2.96) we pick up an extra term, which is negative of the extra term above. Hence, these extra terms cancel in (2.97), and the proof proceeds exactly as before. $\diamond$

Lemma 2.20. *Assume that a positive solution branch $u(r, \lambda)$ loses positivity forward at some $\bar{\lambda}$. Then the transversality condition holds, i.e., the sum of the integrals in (2.93) is not zero.*

Proof: Assume, on the contrary, that $u'_\lambda(1, \bar{\lambda}) = 0$. From the equation for u_λ

$$u''_\lambda + \frac{n-1}{r} u'_\lambda + \lambda f'(u)u_\lambda + f(u) = 0, \tag{2.99}$$

we conclude that $u''_\lambda(1, \bar{\lambda}) = -f(0) > 0$. This, together with $u_\lambda(1, \bar{\lambda}) = u'_\lambda(1, \bar{\lambda}) = 0$, implies that $u_\lambda(r, \bar{\lambda}) > 0$ for r near 1, which is inconsistent with the definition of $\bar{\lambda}$, since solutions must be decreasing in λ near the end-point $r = 1$, before losing positivity forward. $\diamond$

So, one needs to verify that a solution branch loses positivity forward. We saw in the preceding section that this property follows, if we can show that any nontrivial solution of the linearized problem

$$w'' + \frac{n-1}{r} w' + \lambda f'(u)w = 0, \quad r \in (0,1), \quad w'(0) = 0, \ w(1) = 0 \tag{2.100}$$

is of one sign. We have seen above a number of situations when one can prove that $w(r) > 0$.

We are now ready for our main result on symmetry breaking, see P. Korman [84].

Theorem 2.29. *Assume that any non-trivial solution of the linearized problem (2.100) is positive. Assume that the problem (2.92) has a curve of positive solutions, with $u(0, \lambda)$ decreasing, and assume that this curve* *does not continue as a curve of positive solutions for λ large. Then this curve loses positivity forward at some $\lambda = \bar{\lambda}$, and a family of symmetry breaking solutions bifurcates off this curve at the point $(\bar{\lambda}, u(r, \bar{\lambda}))$. The curve of radial solutions $u(r, \lambda)$ continues locally for $\lambda > \bar{\lambda}$, as a curve of radially symmetric sign changing solutions, with the "tangential direction" $u_\lambda = \frac{1}{2\bar{\lambda}} r u_r$. Moreover, an n-parameter family of non-radial solutions bifurcates off this curve at $\lambda = \bar{\lambda}$.*

Proof: By Theorem 2.25, the solution curve loses positivity forward at some $\bar{\lambda}$, and by Lemma 2.18, $w(r) = r u_r - 2\bar{\lambda} u_\lambda$ solves the linearized problem (2.100) at $\lambda = \bar{\lambda}$. By Lemma 2.17, $w(r) \equiv 0$, i.e., $u_\lambda = \frac{1}{2\bar{\lambda}} r u_r$, as claimed. Since by Lemma 2.17 the point $(\bar{\lambda}, u(r, \bar{\lambda}))$ is radially non-degenerate, we see by restricting to the spaces of radial functions and using the Implicit Function Theorem, that the curve of radial solutions continues past the point $(\bar{\lambda}, u(r, \bar{\lambda}))$. Since u_λ is negative near $r = 1$ at this point, it follows that solutions must become negative near the end-point $r = 1$, for $\lambda > \bar{\lambda}$.

Differentiating the equation (2.91) in x_i, we see that the functions $w_i \equiv \frac{x_i}{r} u_r(r, \bar{\lambda})$ are nontrivial solutions of the linearized problem

$$\Delta w + \lambda f'(u) w = 0 \quad \text{for } |x| < 1, \quad w = 0 \text{ on } |x| = 1. \tag{2.101}$$

By the result of G. Cerami [32], the null-space of (2.101) is either n-dimensional or $n + 1$ dimensional, with one of its elements being a radial function (a similar result was proved by J. Smoller and A. Wasserman, see [176]). In view of Lemma 2.17, the first alternative holds, and so w_i, $i = 1, 2, \ldots, n$, span the null-space of the linearized problem (2.101). It follows that the null-space is invariant under the action of the orthogonal group $O(n)$. We denote by $U = U(r, \lambda) = U(r)$ the curve of radially symmetric solutions, passing through the point $(\bar{\lambda}, u(r, \bar{\lambda}))$. Set $u(x) = U(r) + v(x)$. From the previous discussion, it follows that

$$U(r, \bar{\lambda}) = \bar{u}(r), \quad U_\lambda(r, \bar{\lambda}) = u_\lambda(r, \bar{\lambda}-). \tag{2.102}$$

We rewrite our equation (2.91) as

$$G(\lambda, v) \equiv \Delta v + \lambda \left[f(U + v) - f(U) \right] = 0, \quad |x| < 1, \tag{2.103}$$
$$v = 0 \text{ for } |x| = 1.$$

The problem (2.103) has a trivial solution for all $\lambda > \bar{\lambda}$. We show that *bifurcation from zero* occurs, by replacing the use of the standard Crandall-Rabinowitz theorem [39] by its generalization to the case of kernels invariant under the action of $O(n)$, due to A. Vanderbauwhede [185]. The Lemma 2.20 verifies the crucial transversality condition. We conclude the existence of an n-parameter family of non-radial solutions. (See also J. Smoller and A. Wasserman [174] for a similar argument.)   ◇

Remarks

(1) Symmetry-breaking bifurcation, off a radially symmetric branch, was studied previously by J. Smoller and A.G. Wasserman, who in [176] present a verification of transversality condition, which they attribute to C. Pospeich. The proof of transversality condition in [176] uses implicitly the same assumption that positivity is lost forward. This assumption can be found in condition (b) on page 282 of [176].

(2) So let us recapitulate. We proved that symmetry breaking occurs, if the positive branch loses its positivity forward. This follows, once we prove that any non-trivial solution of the linearized problem is of one sign, and for that we have a number of results.

Combining the last result with those of the preceding section, we have the following detailed description of the solution curve for concave $f(u)$.

Theorem 2.30. *Assume that the function $f(u) \in C^2(R)$ satisfies (for some $u_1 > 0$)*

$$f(0) < 0, \quad f(u) < 0 \text{ for } u \in (0, u_1), \quad f(u) > 0 \text{ for } u > u_1,$$

$$f''(u) < 0 \quad \text{for all } u > 0,$$

$$\lim_{u \to \infty} \frac{f(u)}{u} = 0.$$

Then the problem (2.91) has a curve of positive radially symmetric solutions. This curve makes exactly one turn at some $\lambda = \lambda_0$. For $\lambda > \lambda_0$ the curve has two branches, the upper one which continues without any turns for all $\lambda > \lambda_0$, and the lower one which continues without turns, until at some finite $\bar{\lambda} > \lambda_0$ it develops a zero slope at the end-point (i.e., $u'(1, \bar{\lambda}) = 0$), and loses positivity (but stays radially symmetric) for $\lambda > \bar{\lambda}$. Moreover, an n-parameter family of symmetry breaking solutions bifurcates off the lower branch at $\lambda = \bar{\lambda}$.

This result provides additional information to the Theorem 2.27. Similarly, a statement on existence of an n-parameter family of symmetry breaking solutions bifurcating off the radial branch, can be added to both Theorems 2.26 and 2.28.

The following result involves superlinear $f(u)$.

Theorem 2.31. *Assume that $f(u) \in C^2(\bar{R}_+)$, and there is some $u_0 > 0$, so that*

$$f(u) < 0 \ \text{for } u \in [0, u_0), \quad f(u) > 0 \ \text{for } u > u_0.$$

Assume that

$$\lim_{u \to \infty} \frac{f(u)}{u^l} = m > 0, \quad \text{where } 1 < l < \frac{n+2}{n-2} \text{ if } n \geq 3, \, 1 < l < \infty \text{ for } n = 1, 2.$$

Assume that any non-trivial solution of the corresponding linearized problem (2.100) is of one sign. Then there is a $\bar{\lambda} > 0$, so that for $\lambda \in (0, \bar{\lambda})$ there is a curve of positive (and hence radial) solutions of (2.91), which continues without any turns for $\lambda \in (0, \bar{\lambda})$, it develops a zero slope at the end-point (i.e., $u'(1, \bar{\lambda}) = 0$), and loses positivity (but stays radially symmetric) for $\lambda > \bar{\lambda}$. Moreover, an n-parameter family of symmetry breaking solutions bifurcates off the lower branch at $\lambda = \bar{\lambda}$.

Proof: Existence of positive solutions for small λ follows from A. Ambrosetti, D. Arcoya and B. Buffoni [9]. By the Theorem 2.1, the curve of positive solutions cannot be continued for all λ (the function $f(u)$ has no stable roots). Hence, after some $\bar{\lambda}$, solutions on the curve must cease to be positive. By the above results, positivity is lost forward, and the symmetry breaking occurs. $\diamond$

Example 1 Consider $f(u) = u^p - a$, with constants $1 < p < \frac{n+2}{n-2}$ for $n > 2$, and $1 < p < \infty$ for $n = 2$, and $a > 0$. The above theorem applies, since by Lemma 2.9, any non-trivial solution of the linearized problem (2.100) is of one sign.

2.10 Special equations

We now turn to the famous problem

$$u'' + \frac{n-1}{r}u' + \lambda e^u = 0, \quad r \in (0, 1), \quad u'(0) = 0, \, u(1) = 0, \qquad (2.104)$$

which is often referred to as Gelfand's equation (or Bratu's equation [23], when $n = 2$). By the Implicit Function Theorem, there is a curve of positive solutions, starting at $(\lambda = 0, u(0) = 0)$. The behavior of this curve depends on the dimension n, according to the following classical theorem of D.D.

Joseph and T.S. Lundgren [75], see also the book by J. Bebernes and D. Eberly [17] for a nice exposition.

Theorem 2.32. *([75])* **1.** *For $n = 1$ and $n = 2$, the solution curve makes exactly one turn, and tends to infinity as $\lambda \to 0$.*
2. *For $3 \leq n \leq 9$, the solution curve makes infinitely many turns, and tends to infinity as $\lambda \to \bar{\lambda}$, for some $\bar{\lambda} > 0$. (I.e., there are infinitely many positive solutions at $\lambda = \bar{\lambda}$, and finitely many at all other λ.)*
3. *For $n \geq 10$, the solution curve does not make any turns, and tends to infinity as $\lambda \to \lambda^*$, for some $\lambda^* > 0$.*

The linearized problem for (2.104) is

$$w'' + \frac{n-1}{r} w' + \lambda e^u w = 0, \quad r \in (0,1), \quad w'(0) = 0, \quad w(1) = 0. \quad (2.105)$$

P. Korman [88] has observed that the following proposition holds, which provides the "tangential direction" at any turn of the solution curve.

Proposition 2.2. *Assume that the problem (2.105) admits a non-trivial solution. Then*

$$w(r) = r u_r + 2. \quad (2.106)$$

Proof: One verifies that the function $z(r) \equiv r u_r + 2$ satisfies the equation in (2.105), and $z'(0) = 0$. By scaling $w(r)$, we achieve $w(0) = z(0)$, and hence $w(r) \equiv z(r)$ by the uniqueness result, see Lemma 2.1. $\diamondsuit$

Similarly, in case $f(u) = (u + \gamma)^p$, we can also write down explicitly solution of the linearized equation (see below). We see that the same two nonlinearities (i.e., $f(u) = e^u$ and $f(u) = (u+\gamma)^p$), which were completely analyzed by D.D. Joseph and T.S. Lundgren [75], using Dynamical Systems, are special for the bifurcation theory approach as well.

More generally, we shall obtain an exact multiplicity result for a class of non-autonomous problems

$$u'' + \frac{n-1}{r} u' + \lambda f(r, u) = 0 \quad \text{for } r < 1, \quad u'(0) = u(1) = 0, \quad (2.107)$$

where $f(r, u) = r^s (u+\gamma)^p$. We study positive solutions. While the result of B. Gidas, W.-M. Ni and L. Nirenberg [63] does not apply here (that result requires that $f_r(r, u) \leq 0$), we have the following simple lemma.

Lemma 2.21. *Any positive solution of (2.107) is a strictly decreasing function, i.e.,*

$$u'(r) < 0 \quad \text{for all } r \in (0,1).$$

Proof: It is clear from the equation (2.107) that a local maximum must occur at any critical point of the solution $u(r)$. $\diamond$

The linearized problem for (2.107) is

$$w'' + \frac{n-1}{r}w' + \lambda f_u(r, u)w = 0 \quad \text{for } r < 1, \quad w'(0) = w(1) = 0. \quad (2.108)$$

Lemma 2.22. *Assume that*

$$p < \frac{n+s}{n-2}. \quad (2.109)$$

Let $u(r)$ be a singular positive solution of (2.107) (i.e., the problem (2.108) admits non-trivial solutions). Then

$$w(r) = ru'(r) + \frac{2+s}{p-1}u(r) + \gamma\frac{2+s}{p-1}, \quad (2.110)$$

and $w(r) > 0$, for all $r \in [0, 1)$.

Proof: Let $v(r) = ru'(r) + \mu u + \alpha$, with the positive constants μ and α to be chosen. One sees that $v(r)$ satisfies

$$v'' + \frac{n-1}{r}v' + \lambda f_u(r, u)v = \lambda\left[-2f + \mu(f_u u - f) + \alpha f_u - rf_r\right] \quad (2.111)$$
$$= \lambda r^s\left[(u+\gamma)^{p-1}(\mu pu + \alpha p) + (u+\gamma)^p(-2 - \mu - s)\right].$$

Setting

$$\alpha = \mu\gamma, \quad (2.112)$$

we rewrite the right hand side of (2.111) as

$$\lambda r^s(u+\gamma)^p(-2 - \mu - s + p\mu). \quad (2.113)$$

The quantity in (2.113) vanishes, provided we set $\mu = \frac{2+s}{p-1}$, and then from (2.112), $\alpha = \gamma\frac{2+s}{p-1}$. It follows that the function

$$v(r) = ru'(r) + \frac{2+s}{p-1}u(r) + \gamma\frac{2+s}{p-1}$$

satisfies the linearized equation in (2.108).

We have $v'(0) = w'(0) = 0$, and since $w(r)$ is defined up to a constant factor, we can achieve $w(0) = v(0)$, by scaling $w(r)$. By Lemma 2.1, it follows that $w(r) \equiv v(r)$.

To see that $w(r)$ is positive, it suffices to show that it is a decreasing function, since $w(1) = 0$. In view of the condition (2.109), we have

$$w'(r) = ru''(r) + \frac{p+1+s}{p-1}u'(r) \leq r(u''(r) + \frac{n-1}{r}u'(r)) = -\lambda rf(r, u) < 0,$$

completing the proof.

As a consequence, we see that at any singular solution, the Crandall-Rabinowitz Theorem 1.2 applies (since $\int_0^1 f(r,u)w(r)r^{n-1}\,dr > 0$), and a turn to the left occurs (since $f_{uu}(r,u) > 0$).

Since our nonlinearity vanishes when $r = 0$, the well-known a priori estimates of B. Gidas and J. Spruck [64] also do not apply. We shall derive an a priori estimate, by reducing our problem to an equation of Emden-Fowler type.

Lemma 2.23. *Assume that $\lambda \in [\lambda_1, \lambda_2]$, for some $0 < \lambda_1 < \lambda_2$, and the condition (2.109) holds. Then there is a constant $c > 0$, so that any positive solution of the problem (2.107) satisfies*

$$u(r) < c \quad \text{for all } r \in [0,1), \text{ and all } \lambda \in [\lambda_1, \lambda_2].$$

Proof: In case $n > 2$, we set $t = r^{2-n}$ and $v = u + \gamma$, transforming (2.107) into (with $v = v(t)$)

$$v'' + a(t)v^p = 0, \quad t \in (1, \infty) \tag{2.114}$$
$$v(1) = \gamma, \quad v'(\infty) = 0$$
$$v'(t) > 0, \quad t \in (1, \infty),$$

where

$$a(t) = \lambda \frac{1}{(n-2)^2} t^\alpha, \quad \text{and} \quad \alpha = -2 - \frac{s+2}{n-2}.$$

This is a well-studied equation of Emden-Fowler type. Since $\alpha + p + 1 < 0$, it follows by the Theorem 6 on p. 152 in R. Bellman [18], that $\lim_{t\to\infty} v(t) = v(\infty)$ is finite, implying that $u(0) = \max u(r)$ is bounded, see P. Korman [88] for more details, and for the proof in case $n = 2$. $\diamond$

With all of the ingredients in place, we follow a familiar by now scheme to prove the following exact multiplicity result.

Theorem 2.33. *([88]) Assume that $s > 0$ and $\gamma > 0$ are constants, and $p > 1$ is any constant in case $n = 2$, and in case $n > 2$ it satisfies $p < \frac{n+s}{n-2}$. Then there is a critical λ_0, so that the problem (2.107) has exactly 2, 1 or 0 positive radial solutions, depending on whether $\lambda < \lambda_0$, $\lambda = \lambda_0$ or $\lambda > \lambda_0$. Moreover, all solutions lie on a unique smooth solution curve. This curve starts at $(\lambda = 0, u = 0)$, it makes exactly one turn at λ_0, and then it tends to infinity as $\lambda \downarrow 0$.*

Notice that in case $s > 2$, this theorem allows some supercritical values of $p > \frac{n+2}{n-2}$. We see that the r^s term in the problem (2.107) compensates for the singular behavior at $r = 0$, which happens for supercritical equations. We suspect that the condition (2.109) can be improved to read $p < \frac{n+2+s}{n-2}$.

Finally, we mention that Z. Guo and J. Wei [66] have proved that the solution curve for the singular problem

$$u'' + \frac{n-1}{r}u' + \lambda \frac{r^s}{(1-u)^p} = 0, \quad r \in (0,1), \quad u'(0) = 0, \; u(1) = 0,$$

has (under some conditions) infinitely many turns. Similarly to Lemma 2.22, we can see that at these turning points, solution of the corresponding linearized equation is

$$w(r) = ru'(r) - \frac{2+s}{p+1}u(r) + \frac{2+s}{p+1}.$$

2.11 Oscillations of the solution curve

We saw in the previous sections that complexity of the solution curve of the problem

$$u'' + \frac{n-1}{r}u' + \lambda f(u) = 0, \quad r \in (0,1), \quad u'(0) = 0, \; u(1) = 0, \qquad (2.115)$$

tends to mirror that of the nonlinearity $f(u)$, particularly so in case of dimensions $n = 1$ and $n = 2$. For example, if $f(u)$ is convex, $f(u)$ can have at most one critical point, and correspondingly, the solution curve of (2.115) will usually admit at most one turn, in the dimensions $n = 1$ and $n = 2$. Similarly, for some functions with one inflection point, the solution curve admits either exactly one, or exactly two turns, as we saw for cubics above, and more such examples, in case $n = 1$, will be considered in the next chapter. In case when $f(u)$ changes concavity infinitely many times, the solution curve may have infinitely many turns. This possibility was considered in a number of papers, see e.g., R. Schaaf and K. Schmitt [162], [163], D. Costa, H. Jeggle, R. Schaaf and K. Schmitt [37], H. Kielhoffer and S. Maier [76]. In particular, in D. Costa et al [37], the following resonant boundary value problem was considered

$$Lu + \lambda_1 u + g(u) = h(x) \text{ in } D$$
$$u = 0 \text{ on } \partial D,$$

where $D \subset R^n$, $n \geq 2$, L is a second order self-adjoint uniformly elliptic operator, and λ_1 the principal eigenvalue of $-L$ on D with zero boundary conditions. For certain types of bounded domains D, it was shown that the problem has infinitely many positive solutions, bifurcating from infinity. The condition imposed on the domain D turned out to be somewhat restrictive, since for the important special case, on a unit ball,

$$\Delta u + \lambda_1 u + \sin u = 0 \ \text{ in } |x| < 1 \qquad (2.116)$$
$$u = 0 \ \text{ on } |x| = 1,$$

their result applies only for dimensions $n = 1, 2$.

In this section we describe the result of A. Galstian, P. Korman and Y. Li [60] on the problem

$$\Delta u + \lambda(u + a \sin u) = 0 \ \text{ in } B \qquad (2.117)$$
$$u > 0 \ \text{ in } B$$
$$u = 0 \ \text{ on } B,$$

where B is the unit ball in R^n, $n \geq 1$, and a is a positive constant. The advantage of working with the problem (2.117) is that we get more detailed results, by using the bifurcation theory approach. In particular, we know by our previous results that all positive solutions lie on a unique smooth curve of solutions, whose one end bifurcates from zero, and the other one from infinity. (For the problem (2.116), a continuum of solutions bifurcating from infinity was considered in [37].) The question is: how many times does this curve cross the line $\lambda = \lambda_1$? We will show that for $1 \leq n \leq 5$, the problem (2.117) has infinitely many positive solutions at $\lambda = \lambda_1$, while at any other value of λ, the number of positive solutions is at most finite. If $n \geq 6$, the problem (2.117) has at most finitely many positive solutions for all λ. In case $a = \frac{1}{\lambda_1}$, we see that the problem (2.116) has infinitely many positive solutions, when $1 \leq n \leq 5$, and finitely many when $n \geq 6$, thus extending the result of [37]. What is interesting here, is that while the result depends on the space dimension n, this seems not to be connected with Sobolev's embedding.

Let λ_1 denote the principal eigenvalue of the Laplace operator $-\Delta$ on B, with zero boundary condition, and $\varphi_1(r)$ the corresponding eigenfunction.

Theorem 2.34. *([60]) (i) If $1 \leq n \leq 5$ then, for any $a > 0$, the problem (2.117) has infinitely many positive solutions at $\lambda = \lambda_1$, while at any other value of λ, the number of positive solutions is at most finite. Moreover, all* *solutions lie on a unique smooth curve, whose one end bifurcates from zero at $\lambda = \lambda_1/(1+a)$, and the other one from infinity at $\lambda = \lambda_1$.*

(ii) *If $n \geq 6$ then, for any $a > 0$, the problem (2.117) has only finitely many positive solutions for all λ. All solutions lie on a unique smooth curve, whose one end bifurcates from zero at $\lambda = \lambda_1/(1+a)$, and the other one from infinity at $\lambda = \lambda_1$.*

Observe that in case $a > \pi$, the function $f(u) = u + a\sin u$ can be negative on a finite number of sub-intervals, and the solution curve breaks into a finite number of pieces. We outline the proof next. By the classical theorem of B. Gidas, W.-M. Ni and L. Nirenberg [63], any solution of the problem (2.117) is radially symmetric. By the theory of bifurcation from infinity, see e.g., [194], large solutions approach a constant multiple of $\varphi_1(r)$. This allows us to express $\lambda - \lambda_1$ as a certain oscillatory integral. We study oscillations of this integral about zero, by using some tools from Complex Analysis, which we review in the next sub-section.

2.11.1 *Asymptotics of some oscillatory integrals*

The following two lemmas are taken from [172]. We present their short proofs for completeness.

Lemma 2.24. *Let $f(x) \in C^2[0, a]$ and $\alpha \neq 0$. Then, if $\mu \to \infty$,*

$$\Phi(\mu) \equiv \int_0^a f(x)e^{(i/2)\alpha\mu x^2}dx = \frac{1}{2}\sqrt{\frac{2\pi}{|\alpha|\mu}}e^{i(\pi/4)\delta(\alpha)}f(0) + O(\frac{1}{\mu}), \quad (2.118)$$

where $\delta(\alpha) = \operatorname{sgn}\alpha$.

Proof: We take $\alpha > 0$ for definiteness, and we assume first that $f(x) \equiv 1$. After a change of variable $\sqrt{\alpha\mu}\, x = t$, we obtain

$$\int_0^a e^{(i/2)\alpha\mu x^2}dx = \frac{1}{\sqrt{\alpha\mu}}\left[\int_0^\infty e^{it^2/2}dt - \int_{a\sqrt{\alpha\mu}}^\infty e^{it^2/2}dt\right].$$

The first integral is a Fresnel's integral, and it is equal to $\frac{1}{2}e^{i\pi/4}\sqrt{2\pi}$. In the second integral, we denote $y = a\sqrt{\alpha\mu}$, and integrate by parts

$$\int_y^\infty e^{it^2/2}dt = \int_y^\infty \frac{1}{2it}d\left(e^{it^2}\right) = -\frac{e^{iy^2}}{2yi} + \frac{1}{2i}\int_y^\infty \frac{e^{it^2/2}}{t^2}\,dt.$$

Both terms on the right are $O(\frac{1}{y}) = O\left(\frac{1}{\sqrt{\mu}}\right)$, and so we obtain

$$\int_0^a e^{(i/2)\alpha\mu x^2}dx = \frac{1}{2}e^{i\pi/4}\sqrt{\frac{2\pi}{\alpha\mu}} + O\left(\frac{1}{\mu}\right). \quad (2.119)$$

In case $\alpha < 0$, we make a change of variables $-\sqrt{|\alpha|\mu}\, x = t$, use the Fresnel's integral $\int_0^\infty e^{-it^2/2}dt = \frac{1}{2}e^{-i\pi/4}\sqrt{2\pi}$, and arrive at a formula similar to (2.119), with a minus in the exponent. Combining, we have for any α

$$\int_0^a e^{(i/2)\alpha\mu x^2}dx = \frac{1}{2}e^{i\frac{\pi}{4}\delta(\alpha)}\sqrt{\frac{2\pi}{\alpha\mu}} + O\left(\frac{1}{\mu}\right).$$

For an arbitrary function $f(x)$, we may write

$$f(x) = f(0) + [f(x) - f(0)] = f(0) + xh(x),$$

where $h(x) = \frac{f(x)-f(0)}{x} \in C^1[0,a]$. Then

$$\Phi(\mu) = \frac{1}{2}f(0)\sqrt{\frac{2\pi}{\alpha\mu}}e^{i\frac{\pi}{4}\delta(\alpha)} + \int_0^a e^{(i/2)\alpha\mu x^2}xh(x)\,dx + O\left(\frac{1}{\mu}\right). \quad (2.120)$$

Integrating by parts, similarly to the above, it is easy to see that

$$\int_0^a e^{(i/2)\alpha\mu x^2}xh(x)\,dx = O\left(\frac{1}{\mu}\right) \quad (\mu \to +\infty). \quad (2.121)$$

(Here is an intuitive justification of (2.121). We have fast oscillations in the integral, except for a small interval near $x = 0$. The length of this interval is $O\left(\frac{1}{\sqrt{\mu}}\right)$, and the same asymptotic formula holds for $xh(x)$.) Combining (2.121) with (2.120), we obtain (2.118). $\diamond$

Lemma 2.25. *Assume that the functions $f(x)$ and $g(x) > 0$ are of class $C^2[0,1]$, and satisfy*

$$g'(x) < 0 \quad \text{for all} \quad x \in (0,1], \quad \text{and} \quad g'(0) = 0, \ g''(0) < 0.$$

Then, as $\mu \to \infty$,

$$\int_0^1 f(x)e^{i\mu g(x)}\,dx = e^{i(\mu g(0)-\frac{\pi}{4})}\sqrt{\frac{\pi}{\mu|g''(0)|}}f(0) + O\left(\frac{1}{\mu}\right).$$

Proof: On the interval $[0,1]$, we change the variables $x \to t$, $x = \psi(t)$, so that

$$g(x) - g(0) = -t^2. \quad (2.122)$$

Since $g(x)$ is a decreasing function, g^{-1} exists. Thus

$$\psi(t) = g^{-1}(g(0) - t^2).$$

Note, that

if $x = 0$, then $t = 0$; if $x = 1$ then $t = \sqrt{g(0) - g(1)} \equiv b > 0$.

Therefore, after the change of variables, we obtain

$$I = \int_0^1 f(x)e^{i\mu g(x)}\,dx = e^{i\mu g(0)}\int_0^b f(\psi(t))\psi'(t)e^{-i\mu t^2}\,dt\,.$$

To apply Lemma 2.24, we need to calculate $\psi'(0)$. Differentiating (2.122), we have

$$g'(\psi(t))\psi'(t) = -2t\,.$$

We cannot calculate $\psi'(0)$ from here, since $g'(0) = 0$. So, we differentiate this formula again, and set $t = 0$, obtaining

$$\psi'(0) = \sqrt{-\frac{2}{g''(0)}}\,.$$

Applying Lemma 2.24 to the integral I, we conclude the proof. $\diamond$

Corollary Under the conditions of Lemma 2.25, as $\mu \to \infty$, we have

$$\int_0^1 f(x)e^{i\mu g(x)}\,dx = O\left(\frac{1}{\sqrt{\mu}}\right)\,.$$

2.11.2 *Reduction to the oscillatory integrals*

Our first lemma shows that the positive solutions of (2.117) lie in a strip, whose projection on the λ axis is bounded.

Lemma 2.26. *If the problem (2.117) has a positive solution, then*

$$\frac{\lambda_1}{2} < \lambda < \frac{\pi}{\pi - 1}\lambda_1\,. \tag{2.123}$$

Proof: Multiplying both sides of the equation in (2.117) by u, and integrating, we obtain:

$$\int_B u\Delta u\,dx + \lambda\int_B (u^2 + u\sin u)\,dx = 0\,.$$

Since $u^2 + u\sin u \leq 2u^2$, after integration by parts, and using the Poincare inequality, we have

$$2\lambda\int_B u^2\,dx > \lambda\int_B (u^2 + u\sin u)\,dx = \int_B |\nabla u|^2 dx \geq \lambda_1\int_B u^2\,dx\,,$$

from which the left hand side of the inequality (2.123) follows. Multiplying the equation in (2.117) by φ_1, and integrating, we obtain

$$\int_B [\lambda(u + \sin u) - \lambda_1 u]\varphi_1\,dx = 0\,. \tag{2.124}$$

The second part of the inequality (2.123) we prove by contradiction. Assume, on the contrary, that at some solution

$$\lambda \geq \frac{\pi}{\pi - 1} \lambda_1 > \lambda_1 \,. \tag{2.125}$$

Denote $g(u) = \frac{\lambda - \lambda_1}{\lambda} u + \sin u$. From (2.125) we obtain immediately, that $g(u) > 0$ on $(0, \pi)$. Note that $g(\pi) = \frac{\lambda - \lambda_1}{\lambda} \pi$. It is clear, that if $g(\pi) \geq 1$, then $g(u)$ is positive for all $u > 0$. But from (2.125) we have

$$\lambda(\pi - 1) \geq \pi \lambda_1 \implies \pi(\lambda - \lambda_1) \geq \lambda \implies \frac{\pi(\lambda - \lambda_1)}{\lambda} \geq 1 \,.$$

Thus $g(u)$ is positive. Since $\varphi_1 > 0$, we obtain a contradiction in (2.124). Therefore, $\lambda < \frac{\pi}{\pi - 1} \lambda_1$. $\diamondsuit$

For the rest of this section we shall assume that $a = 1$, for simplicity. There is a curve of positive solutions of (2.117) bifurcating off the trivial solution $u = 0$ (to the right) at $\frac{\lambda_1}{2}$. (This follows by a standard result of M.G. Crandall and P.H. Rabinowitz [39].) Similarly, there is a curve of positive solutions bifurcating from infinity at $\lambda = \lambda_1$, see. e.g., p. 673 in E. Zeidler [194]. Let us consider the branch bifurcating from infinity. This branch extends globally, since at each point either the implicit function theorem or a bifurcation Theorem 1.2 applies. By Lemma 2.26, this branch is constrained to a strip $\frac{\lambda_1}{2} < \lambda < \frac{\pi}{\pi - 1} \lambda_1$, so it has to go to zero at $\lambda_1/2$. By the uniqueness of the solution, bifurcating from zero, this branch has to link up with the lower one. We thus obtain a solution curve, connecting $(\lambda_1/2, 0)$ to (λ_1, ∞), which exhausts the set of all possible solutions of (2.117), by Lemma 2.2.

Since we are interested in the behavior of solution $u(x)$ near infinity in L^∞, it is appropriate to use the scaling

$$u = \mu v, \quad \text{so that} \quad |v|_{L^\infty} \equiv 1, \quad \text{while} \quad \mu \to \infty \,.$$

Then (2.117) becomes

$$\Delta v + \lambda v + \frac{\lambda}{\mu} \sin(\mu v) = 0 \text{ in } B, \ v = 0 \text{ on } \partial B \,. \tag{2.126}$$

Multiplying (2.126) by φ_1, and integrating by parts, we obtain

$$\lambda - \lambda_1 = -\frac{\lambda}{\mu} \frac{\int_B \sin(\mu v) \varphi_1 \, dx}{\int_B v \varphi_1 \, dx} \,. \tag{2.127}$$

By the classical theorem of B. Gidas, W.-M. Ni and L. Nirenberg [63], any solution of the problem (2.117) is radially symmetric. Therefore, the problem (2.126) becomes:

$$v'' + \frac{n - 1}{r} v' + \lambda v + \frac{\lambda}{\mu} \sin(\mu v) = 0, \ 0 < r < 1, \ v'(0) = v(1) = 0 \,. \tag{2.128}$$

By the theory of bifurcation from infinity, $v \to \varphi_1$ in the norm C^2, as $\mu \to \infty$, see e.g., [194]. Hence, the integral $\int_B v\varphi_1 dx$ is positive for large μ. So, the issue is whether the integral $\int_B \sin(\mu v)\varphi_1 dx$ changes sign infinitely many or finitely many times, as $\mu \to \infty$. Correspondingly, the solution curve will cross the line $\lambda = \lambda_1$ either infinitely many or finitely many times, as follows from (2.127).

We have:

$$\int_B \sin(\mu v)\varphi_1 \, dx = \omega_n \int_0^1 \varphi_1(r) \sin(\mu v(r)) r^{n-1} \, dr$$
$$= \omega_n \text{Im} \int_0^1 \varphi_1(r) r^{n-1} e^{i\mu v(r)} \, dr,$$

where ω_n is the area of the unit sphere in R^n, and Im denotes the imaginary part. We shall study the integral

$$I \equiv \text{Im} \int_0^1 \varphi_1(r) r^{n-1} e^{i\mu v(r)} \, dr. \tag{2.129}$$

Depending on the dimension n, we will need up to three integrations by parts in (2.129). To spare the reader, we do only one integration by parts here, and the rest can be found in [60]. With $f_1(r) \equiv \frac{\varphi_1(r) r^{n-1}}{v'(r)}$, we have

$$\int_0^1 \varphi_1(r) r^{n-1} e^{i\mu v(r)} \, dr = \frac{1}{i\mu} \int_0^1 \frac{\varphi_1(r) r^{n-1}}{v'} d(e^{i\mu v(r)})$$
$$= \frac{1}{i\mu} \int_0^1 f_1(r) d(e^{i\mu v(r)}) = \frac{1}{i\mu} f_1(r) e^{i\mu v(r)} \Big|_0^1$$
$$- \frac{1}{i\mu} \int_0^1 f_1'(r) e^{i\mu v(r)} \, dr. \tag{2.130}$$

Sketch of proof of Theorem 2.34 The case of $n = 1$ has already been covered in [104], so we start with $n = 2$.

(i) $n = 2$. We use (2.130). Here $f_1(r) = \frac{r\varphi_1(r)}{v'(r)}$. Observe that $f_1(1) = 0$, $f_1(0) = \frac{\varphi_1(0)}{v''(0)} \simeq \frac{\varphi_1(0)}{\varphi_1''(0)} < 0$ for large μ. Using the Corollary of Lemma 2.25, we obtain:

$$I = \text{Im} \frac{1}{i\mu} \left[-f_1(0) e^{i\mu v(0)} - \int_0^1 f_1'(r) e^{i\mu v(r)} \, dr \right]$$
$$= \frac{f_1(0)}{\mu} \cos \mu v(0) + O\left(\frac{1}{\mu^{3/2}}\right), \quad \text{as} \quad \mu \to \infty.$$

It follows that I changes sign infinitely many times, as $\mu \to \infty$. (Here, of course, $v(0) = 1$.)

(ii) $n = 3$. We use (2.130) again. Here $f_1(r) = \frac{r^2 \varphi_1(r)}{v'(r)}$. This time we have $f_1(0) = f_1(1) = 0$, while by the L'Hospital rule, $f_1'(0) \neq 0$. Hence, by Lemma 2.25,

$$I = \operatorname{Im} \left[\frac{i}{\mu} \int_0^1 f_1'(r) e^{i\mu v(r)} dr \right] = \operatorname{Re} \left[\frac{c_1}{\mu^{3/2}} f_1'(0) e^{i(\mu v(0) - \frac{\pi}{4})} \right] + O\left(\frac{1}{\mu^2} \right),$$

where $c_1 = \sqrt{\frac{\pi}{|v''(0)|}} \simeq \sqrt{\frac{\pi}{|\varphi_1''(0)|}}$. As above, we see that I changes sign infinitely many times as $\mu \to \infty$.

The rest of proof involves considering separately the dimensions $n = 4$, $n = 5$, $n = 6$, and $n \geq 7$. Depending on the dimension, one needs to perform either one more, or two more integrations by parts in (2.130), see [60]. $\diamond$

2.12 Uniqueness for non-autonomous problems

We study uniqueness of positive solution for a class of non-autonomous problems on a ball

$$\Delta u + q(|x|)u + u^p = 0 \quad \text{for } |x| < R, \ u = 0, \text{ when } |x| = R. \quad (2.131)$$

Here $x \in R^n$, $u = u(x)$, and $q = q(r)$, $r = |x|$, is a given differentiable function. The constant p is assumed to be sub-critical, $1 < p \leq \frac{n+2}{n-2}$. This problem was studied by M.K. Kwong and Y. Li [121]. In Theorem 2 of that paper, a condition for uniqueness of positive radial solution is given. Namely, they define the constants $\beta = \frac{2(n-1)(p-1)}{p+3}$ and $L = \frac{2(n-1)(np+n-2p-4)}{(p+3)^2}$, and show that the problem (2.131) has at most one positive solution, provided that $1 < p < \frac{n+2}{n-2}$, $n \geq 3$, and the function

$$G(r) := r^\beta q(r) - L r^{\beta-2} \quad \text{has a } \wedge \text{ property}, \quad (2.132)$$

i.e., there exists some $c \in [0, R]$, so that the function $G(r)$ is increasing on $[0, c)$, and decreasing on $(c, R]$ (notice that both increasing and decreasing $G(r)$ are included in this definition). Since the condition (2.132) can be cumbersome to verify, M.K. Kwong and Y. Li [121] had provided the following corollary: condition (2.132) can be replaced by

$$r^\beta q(r) \quad \text{is a nondecreasing function.} \quad (2.133)$$

By adapting a technique of Adimurthy, F. Pacella and S.L. Yadava [4], P. Korman [85] proved uniqueness of positive solution of (2.131) for $n \geq 1$, $1 < p \leq p^*$, where $p^* = \frac{n+2}{n-2}$ for $n \geq 3$, $p^* = \infty$ for $n = 1, 2$; and

$$\frac{d}{dr} \left[r^{2c_n} q(r) \right] \geq 0, \quad (2.134)$$

where $c_n = n(\frac{1}{2} - \frac{1}{p+1})$. We shall present this result in this section. For differentiable $q(r)$, our condition (2.134) improves the condition (2.133) for all dimensions n, as one easily checks that inequality $2c_n > \beta$ is equivalent to p being sub-critical. Our condition (2.134) also provides an extension of the full uniqueness result in M.K. Kwong and Y. Li [121], as the following example shows.

Example. Take $n = 4$, $p = 2$ and $R = 50$. Then one has $2c_4 = 4/3 > \beta = 6/5$. Compute also $L = 24/25$ and $\beta - 2 = -4/5$. Assume that

$$q(r) = 0.02 + \frac{20}{1 + r^{4/3}}.$$

Then $r^{4/3}q(r)$ is increasing so that our condition (2.134) holds. On the other hand,

$$r^\beta q(r) - Lr^{\beta-2} = 0.02r^{6/5} + \frac{20r^{6/5}}{1 + r^{4/3}} - \frac{24}{25}r^{-4/5}$$

is not of type $\wedge$ on $(0, 50)$.

We remark that the approaches of M.K. Kwong and Y. Li [121], and of Adimurthy, F. Pacella and S.L. Yadava [4] are similar. Both papers show that any two positive solutions cannot intersect, which is done by examining the ratio of the solutions. Adimurthy, F. Pacella and S.L. Yadava [4] use a Pohozhaev identity to complete the argument, whereas M.K. Kwong and Y. Li [121] were using the energy function.

We shall consider positive radially symmetric solutions for a little more general problem on a ball $|x| < R$ in R^n

$$\Delta u + f(r, u) = 0 \quad \text{for } |x| < R, \quad u = 0 \quad \text{on } |x| = R. \tag{2.135}$$

We set $F(r, u) = \int_0^u f(r, s)\, ds$, and introduce the function $H(r) = \frac{1}{2}ru_r^2 + \alpha u u_r + r F(r, u)$, with a constant α to be chosen shortly. Then, using the equation (2.135), one easily verifies that

$$H' + \frac{n-1}{r}H \tag{2.136}$$
$$= \left(1 - \frac{n}{2} + \alpha\right)u_r^2 + nF(r, u) - \alpha u f(u, r) + r F_r(r, u).$$

We now return to the problem

$$u'' + \frac{n-1}{r}u' + q(r)u + u^p = 0 \quad \text{for } r < R, \quad u'(0) = u(R) = 0. \tag{2.137}$$

We assume that $q(r)$ is a positive function of class $C^1(0, R) \cap C[0, R]$, and p is a sub-critical power, $1 < p \le p^*$, where $p^* = \frac{n+2}{n-2}$, if $n > 2$ and $p^* = \infty$

for $n = 1, 2$. For this problem it is convenient to select $\alpha = \frac{n}{p+1}$. Then (2.136) takes the form

$$H' + \frac{n-1}{r}H = (1 - c_n)u_r^2 + c_n q u^2 + \frac{1}{2}r q' u^2, \qquad (2.138)$$

where we denote $c_n = n(\frac{1}{2} - \frac{1}{p+1})$. Notice that the condition, $1 < p \leq \frac{n+2}{n-2}$ for $n \geq 3$, is equivalent to

$$0 < c_n \leq 1, \qquad (2.139)$$

and in the cases $n = 1, 2$, (2.139) holds for any $1 < p < \infty$.

Multiplying the relation (2.138) by an integrating factor r^{n-1}, and integrating over $[r_1, r_2] \subset (0, 1)$, we obtain a generalized Pohozhaev identity

$$\int_{r_1}^{r_2} \left[(1 - c_n)u_r^2 + c_n q u^2 + \frac{1}{2}r q' u^2 \right] r^{n-1}\, dr \qquad (2.140)$$
$$= r_2^{n-1}H(r_2) - r_1^{n-1}H(r_1).$$

Notice that the quantity on the left in (2.140) is positive, provided that $rq' + 2c_n q > 0$, i.e., if

$$\frac{d}{dr}\left[r^{2c_n}q(r) \right] \geq 0, \qquad (2.141)$$

and in case $c_n = 1$ (i.e., when $p = p^*$, and $n \geq 3$), the above inequality is assumed to be strict on a set of positive measure.

We record the following simple observation.

Lemma 2.27. *Assume that $q(r) > 0$ for all $r \in (0, R)$. Then any positive solution of (2.137) satisfies*

$$u'(r) < 0 \quad \text{for all } r \in (0, R).$$

Proof: From the equation (2.137), any critical point of the solution inside $(0, R)$ is a point of local maximum, and hence $r = 0$ is the only critical point. $\diamond$

Theorem 2.35. *Assume that $0 < q(r) \in C^1(0, R) \cap C[0, R]$ satisfies the condition (2.141), and $p \leq p^*$ is sub-critical. In case when $p = \frac{n+2}{n-2}$, and $n \geq 3$, we assume additionally that the inequality (2.141) is strict on a set of positive measure. Then any two positive solutions of (2.137) do not intersect (i.e., they are strictly ordered on $(0, R)$).*

Proof: The proof is modeled after the one in Adimurthy et al [4]. Let $v_1(r)$ and $v_2(r)$ be two positive solutions of (2.137), and we may assume that $v_1(0) < v_2(0)$, by Lemma 2.1. Assume, on the contrary, that the solutions intersect. It follows that the function $z(r) \equiv \frac{v_1(r)}{v_2(r)}$ becomes greater than 1 somewhere on $(0, R)$, while $z(0) < 1$. Let $\xi_0 \in (0, R]$ be the first local maximum of $z(r)$ with $z(\xi_0) > 1$, and denote $t_0 = z(\xi_0)$. If $\xi_0 = R$, then

$$t_0 = \frac{v_1'(R)}{v_2'(R)} . \tag{2.142}$$

If $\xi_0 \in (0, R)$, then $z'(\xi_0) = 0$. In both cases, we have the following two inequalities

$$1 < t_0 = \frac{v_1(\xi_0)}{v_2(\xi_0)} = \frac{v_1'(\xi_0)}{v_2'(\xi_0)} , \tag{2.143}$$

$$v_1(r) < t_0 v_2(r) \quad \text{for all } r \in [0, \xi_0) . \tag{2.144}$$

We claim that

$$\frac{v_1'(r)}{v_2'(r)} < t_0 \quad \text{for all } r \in [0, \xi_0) . \tag{2.145}$$

Denote $p(r) = v_1'v_2 - v_1v_2'$. Then, from the equation (2.137), written for both v_1 and v_2, we obtain

$$\frac{d}{dr}\left(r^{n-1}p(r)\right) + r^{n-1}v_1v_2\left(v_1^{p-1} - v_2^{p-1}\right) = 0 . \tag{2.146}$$

From the definition of ξ_0, as the *first* point of maximum of $z(r)$ (with maximum value above one), it follows that $v_1(r)$ and $v_2(r)$ intersect exactly once on $(0, \xi_0)$. Let $r_1 < \xi_0$ be such that $v_1(r_1) = v_2(r_1)$. Then, the function $r^{n-1}p(r)$ vanishes at $r = 0$ and $r = \xi_0$, it is increasing on $(0, r_1)$, and decreasing on (r_1, ξ_0). It follows that $p(r)$ is positive on $(0, \xi_0)$, i.e., using Lemma 2.27,

$$\frac{v_1'(r)}{v_2'(r)} < \frac{v_1(r)}{v_2(r)} \quad \text{for all } r \in [0, \xi_0) . \tag{2.147}$$

From (2.147) and (2.144), the claim (2.145) follows.

We now distinguish between two cases.

Case 1. $\xi_0 = R$. Taking a linear combination of two Pohozhaev identities (2.140), the ones written for $v_1(r)$ and $v_2(r)$ respectively, we have

$$(1 - c_n) \int_0^R \left(v_1'^2 - t_0^2 v_2'^2\right) r^{n-1}\,dr + \int_0^R Q(r)\left(v_1^2 - t_0^2 v_2^2\right) r^{n-1}\,dr \tag{2.148}$$
$$= \tfrac{1}{2}\left(v_1'^2(R) - t_0^2 v_2'^2(R)\right) R^n,$$

where $Q(r) = \frac{d}{dr}\left[r^{2c_n}q(r)\right] \geq 0$. In view of (2.142), the quantity on the right is zero, while the one on the left is negative, by (2.144) and (2.145), a contradiction.

Case 2. $\xi_0 < R$. Similarly to the above, we have

$$(1 - c_n)\int_0^{\xi_0}\left(v_1'^2 - t_0^2 v_2'^2\right)r^{n-1}\,dr + \int_0^{\xi_0} Q(r)\left(v_1^2 - t_0^2 v_2^2\right)r^{n-1}\,dr \quad (2.149)$$
$$= \xi_0^{n-1}\left[H_{v_1}(\xi_0) - t_0^2 H_{v_2}(\xi_0)\right],$$

where by $H_{v_1}(\xi_0)$ we denote the function $H(r)$ defined above, with $v_1(r)$ replacing $u(r)$. The quantity on the left is negative, as before. Using the definition of t_0 and the formula (2.143), we see that two pairs of terms cancel on the right. What remains on the right is

$$\xi_0^{n-1}\left(H_{v_1}(\xi_0) - t_0^2 H_{v_2}(\xi_0)\right) = \xi_0^n\left[F(v_1, r) - t_0^2 F(v_2, r)\right]$$
$$= \frac{\xi_0^n}{(p+1)}\left(v_1^{p+1}(\xi_0) - \frac{v_1^2(\xi_0)}{v_2^2(\xi_0)}v_2^{p+1}(\xi_0)\right) > 0,$$

since $v_1(\xi_0) > v_2(\xi_0)$. We have a contradiction. It follows that $v_1(r)$ and $v_2(r)$ cannot intersect. $\diamond$

The uniqueness result now follows easily.

Theorem 2.36. *In the conditions of the preceding theorem, the problem (2.137) has at most one positive radial solution.*

Proof: By the preceding theorem, any two positive solutions $v_1(r)$ and $v_2(r)$ of (2.137) are strictly ordered. Multiplying the equation (2.137), written for $v_1(r)$, by $v_2(r)$, and the equation for $v_2(r)$ by $v_1(r)$, integrating over $(0, R)$ and subtracting, we have

$$\int_0^R v_1 v_2\left[v_1^{p-1} - v_2^{p-1}\right]r^{n-1}\,dr = 0,$$

a contradiction. $\diamond$

Using a different method, P. Korman [93]) had established both existence and uniqueness of positive solutions for the problem (here $r = |x|$)

$$\Delta u - a(r)u + b(r)u^p = 0, \quad \text{for } r < 1, \quad\quad (2.150)$$
$$u = 0 \text{ for } r = 1,$$

and we describe this result next. We shall assume that $p < \frac{n+2}{n-2}$, and the functions $a(r), b(r) \in C^1[0, 1]$ satisfy

$$a(r) > 0, \quad b(r) > 0, \quad a'(r) > 0, \quad b'(r) < 0 \quad \text{for } r \in (0, 1). \quad\quad (2.151)$$

These assumptions imply, in particular, that any positive solution of (2.150) is radially symmetric, in view of B. Gidas, W.-M. Ni and L. Nirenberg [63]. We also define the functions

$$A(r) \equiv 2a(r) + ra'(r),$$

$$B(r) \equiv \left(\frac{n}{p+1} - \frac{n-2}{2} \right) b(r) + \frac{rb'(r)}{p+1}.$$

Observe that we have $\frac{n}{p+1} - \frac{n-2}{2} > 0$ for sub-critical p, i.e., when $p < \frac{n+2}{n-2}$.

Theorem 2.37. *In addition to the conditions (2.151), assume that the function $A(r)$ is positive and increasing, while the function $B(r)$ is positive on $(0,1)$. Assume also that the function $\frac{rb'(r)}{b(r)}$ is decreasing on $(0,1)$. Then the problem (2.150) has a unique positive solution. Moreover, the Morse index of this solution is equal to one.*

We shall present the proof of this theorem, after some preliminary work. We shall consider a more general problem

$$u'' + \frac{n-1}{r} u' + f(r,u) = 0, \quad \text{for all } r \in [0,1], \quad u'(0) = 0, \quad u(1) = 0, \quad (2.152)$$

where $f(r,u)$ satisfies

$$f(r,0) = 0, \quad \text{and} \quad f_r(r,u) \le 0, \quad \text{for all } r \in [0,1], \text{ and } u > 0; \quad (2.153)$$

$$uf_u(r,u) - f(r,u) > 0 \quad \text{for all } r \in [0,1], \text{ and } u > 0. \quad (2.154)$$

In addition, we assume the following condition, which originated from A. Aftalion and F. Pacella [11]

$$\text{the function } \alpha(r) \equiv \frac{2f(r,u(r)) + rf_r(r,u(r))}{u(r)f_u(r,u(r)) - f(r,u(r))} \quad (2.155)$$

is a non-increasing function of r, for $r \in (0,1)$.

We will show that under these conditions any positive solution of (2.152) is non-singular, i.e., the corresponding linearized problem

$$w'' + \frac{n-1}{r} w' + f_u(r,u)w = 0, \quad w'(0) = 0, \quad w(1) = 0 \quad (2.156)$$

admits only the trivial solution $w = 0$. We start with two technical lemmas.

Lemma 2.28. *Let $u(r)$ be a positive solution of (2.152), and assume that the function $f(r,u)$ satisfies the conditions (2.153) and (2.154). Then the function $f(r,u(r))$ can change sign at most once on $(0,1)$.*

Proof: Let $\xi \in (0,1)$ be such that $f(\xi, u(\xi)) = 0$. We claim that $f(r, u(r)) > 0$ for all $r \in [0, \xi)$. Indeed, from (2.154), we conclude that $f_u(\xi, u(\xi)) > 0$, and in general $f_u(r, u(r)) > 0$, so long as $f(r, u(r)) > 0$, and hence

$$\frac{d}{dr} f(r, u(r)) = f_r(r, u(r)) + f_u(r, u(r))u'(r) < 0, \quad \text{for all } r \in [0, \xi),$$

and the claim follows. So that, if the function $f(r, u(r))$ is positive near $r = 1$, it is positive for all $r \in [0, 1)$. If, on the other hand, $f(r, u(r))$ is negative near $r = 1$, it will change sign exactly once on $[0, 1)$ (since it cannot stay negative for all r, by the maximum principle, applied to (2.152)). $\diamond$

The lemma implies that there is a point $r_2 \in (0, 1]$, so that

$$f(r, u(r)) > 0 \quad \text{on } [0, r_2), \quad \text{and } f(r, u(r)) < 0 \text{ on } (r_2, 1). \tag{2.157}$$

(In case $r_2 = 1$, the interval $(r_2, 1)$ is empty.)

Lemma 2.29. *Assume that the conditions of the preceding lemma hold, and in addition,*

$$2f(r, u) + rf_r(r, u) > 0, \quad \text{for all } r \in [0, 1), \text{ and } u > 0. \tag{2.158}$$

Then any solution of the linearized problem (2.156), $w(r)$, cannot vanish in the region where $f(r, u(r)) < 0$ (i.e., on $(r_2, 1)$).

Proof: Assume that the contrary is true, and let τ denote the largest root of $w(r)$, in the region where $f(r, u(r)) < 0$. We may assume that $w(r) > 0$ on $(\tau, 1)$. Then integrating the formula (2.164) below over $(\tau, 1)$, we have

$$u'(1)w'(1) - \tau^n u'(\tau)w'(\tau) = \int_\tau^1 (2f(r, u(r)) + rf_r(r, u(r))) \, wr^{n-1} \, dr.$$

We have a contradiction, since the left hand side is positive, while the integral on the right is negative. $\diamond$

We shall need again the following functions:

$$Q(r) = r^n \left[u'(r)^2 + u(r)f(r, u(r)) \right] + (n-2)r^{n-1}u'(r)u(r), \tag{2.159}$$

$$P(r) = r^n \left[u'(r)^2 + 2F(r, u(r)) \right] + (n-2)r^{n-1}u'(r)u(r), \tag{2.160}$$

where we denote $F(r, u) = \int_0^u f(r, t) \, dt$. Clearly,

$$Q(r) = P(r) + r^n \left[u(r)f(r, u(r)) - 2F(r, u(r)) \right]. \tag{2.161}$$

Observe that $P(0) = 0$, $P(1) = u'(1)^2 > 0$, and

$$P'(r) = r^{n-1} \left[2nF(r, u(r)) - (n-2)u(r)f(r, u(r)) + 2rF_r(r, u(r)) \right]$$
$$\equiv r^{n-1} I(r).$$

The following lemma follows immediately. It gives three sets of conditions for the positivity of $P(r)$.

Lemma 2.30. **(i)** *Assume there is an $0 < r_0 \leq 1$, such that $I(r) > 0$ on the interval $(0, r_0)$. Then $P(r) > 0$ on $(0, r_0]$.*

(ii) *Assume there is an $0 \leq r_0 < 1$, such that $I(r) < 0$ on the interval $(r_0, 1)$. Then $P(r) > 0$ on $[r_0, 1)$.*

(iii) *Assume there is an $0 < r_0 < 1$, such that $I(r) > 0$ on the interval $(0, r_0)$ and $I(r) < 0$ on the interval $(r_0, 1)$. Then $P(r) > 0$ on $(0, 1]$.*

Theorem 2.38. *Assume that the conditions (2.153), (2.154), (2.155) and (2.158) hold. Assume also that either the condition **(i)** of Lemma 2.30 holds with $r_0 = r_2$, or the condition **(ii)** of Lemma 2.30 holds with $r_0 = 0$, or the condition **(iii)** of that lemma holds. Assume finally that*

$$u(r)f(r, u(r)) - 2F(r, u(r)) > 0 \quad \text{for } r \in (0, r_2). \tag{2.162}$$

Then any positive solution of (2.152) is non-singular, i.e., the corresponding linearized problem (2.156) admits only the trivial solution.

Proof: We use again the generalized Wronskians, $\xi(r) = r^{n-1} (u'w - uw')$, and $\zeta(r) = r^n [u'w' + f(r, u)w] + (n-2)r^{n-1}u'w$. We have

$$\xi'(r) = [uf_u(r, u) - f(r, u)] wr^{n-1}, \tag{2.163}$$

$$\zeta'(r) = 2f(r, u)wr^{n-1} + r^n f_r(r, u)w. \tag{2.164}$$

As before, we introduce the function $O(r) = 2\gamma\xi(r) - \zeta(r)$, which in view of (2.163) and (2.164), satisfies

$$O'(r) = 2 [u(r)f_u(r, u(r)) - f(r, u(r))] w(r)r^{n-1} [\gamma - \alpha(r)], \tag{2.165}$$

with $\alpha(r)$ as defined by (2.155).

We may assume that $w(0) > 0$. We claim that the function $w(r)$ cannot have any roots inside $(0, 1)$. Assuming otherwise, let τ_1 be the smallest root of $w(r)$, i.e., $w(r) > 0$ on $[0, \tau_1)$. Let $\tau_2 \in (\tau_1, 1]$ denote the second root of $w(r)$. We now fix $\gamma = \alpha(\tau_1)$. By monotonicity of $\alpha(r)$, the function $\gamma - \alpha(r)$

is non-positive on $[0, \tau_1)$ and non-negative on (τ_1, τ_2). Since $O(0) = 0$, we conclude from (2.165) that

$$O(r) \leq 0 \quad \text{for all } r \in [0, \tau_2]. \tag{2.166}$$

We consider two cases.

Case (i) $\tau_2 = 1$. From (2.166) we have $O(1) \leq 0$. On the other hand, $O(1) = -\zeta(1) = -u'(1)w'(1) > 0$, a contradiction.

Case (ii) $\tau_2 < 1$. Observe that $f(\tau_2, u(\tau_2)) > 0$. Indeed, assuming otherwise, we conclude by Lemma 2.28 that $f(r, u(r)) \leq 0$ on $(\tau_2, 1)$. But this contradicts Lemma 2.29. Applying Lemma 2.28 again, we conclude that $f(r, u(r)) > 0$ over $[0, \tau_2)$.

Since $\xi(\tau_1) > 0$, while $\xi(\tau_2) < 0$, we can find $t \in (\tau_1, \tau_2)$, such that $\xi(t) = 0$, i.e.,

$$\frac{u(t)}{w(t)} = \frac{u'(t)}{w'(t)}. \tag{2.167}$$

By Lemma 2.29, $t \in (0, r_2)$, and then by Lemma 2.30, (2.161) and (2.162),

$$Q(t) > 0. \tag{2.168}$$

In view of (2.166),

$$\zeta(t) = -O(t) \geq 0.$$

On the other hand, using (2.167) and (2.168),

$$\zeta(t) = \left[t^n \left(u'w'\frac{u}{w} + f(t, u)u \right) + (n-2)t^{n-1}u'u \right] \frac{w}{u} = Q(t)\frac{w(t)}{u(t)} < 0,$$

giving us a contradiction.

It follows that $w(r)$ cannot have any roots, i.e., we may assume that $w(r) > 0$ on $[0, 1)$. But that is impossible, as can be seen by integrating (2.163) over $(0, 1)$. Hence $w \equiv 0$.   ◇

Proof of Theorem 2.37 We embed the problem (2.150) into a two-parameter family of problems, with parameters $\theta \in [0, 1]$, and $\lambda \in [0, 1]$,

$$\Delta u - \lambda a(r)u + b^\theta(r)u^p = 0, \quad \text{for } r \in (0, 1), \tag{2.169}$$

$$u = 0 \text{ for } r = 1.$$

At $\lambda = 0$ and $\theta = 0$, the problem has unique solution of Morse index one. This is known for general domains, see e.g., K.C. Chang [33], or a simple proof for the case of a ball can be found above. At $\lambda = 1$ and $\theta = 1$, we have the original problem (2.150).

We check that the Theorem 2.38 applies for all $\theta \in [0,1]$, and all $\lambda \in [0,1]$. Compute

$$(p-1)\alpha(r) = -\lambda \frac{A(r)}{b^\theta(r)u^{p-1}(r)} + 2 + \theta \frac{rb'(r)}{b(r)}.$$

In view of our assumption, $\alpha(r)$ is a decreasing function, thus verifying (2.155). Compute

$$uf(u) - 2F = \frac{p-1}{p+1}b^\theta(r)u^{p+1} > 0,$$

verifying (2.162).

Denote $B_\theta(r) = \left(\frac{n}{p+1} - \frac{n-2}{2}\right)b^\theta(r) + \theta\frac{rb^{\theta-1}(r)b'(r)}{p+1}$. Observe that $B_\theta(r) > b^{\theta-1}(r)B(r) > 0$. Writing $B_\theta(r) = b^\theta(r)\left[\frac{n}{p+1} - \frac{n-2}{2} + \frac{\theta}{p+1}\frac{rb'(r)}{b(r)}\right]$, we see that the quantity in the square bracket is positive and decreasing on $(0,1)$, and hence $B_\theta(r)$ is positive and decreasing on $(0,1)$. Compute

$$I(r) = -\lambda A(r)u^2 + 2B_\theta(r)u^{p+1} = 2B_\theta(r)u^2\left[-\frac{\lambda A(r)}{2B_\theta(r)} + u^{p-1}\right].$$

The quantity in the square bracket is a decreasing function, which is negative near $r = 1$. Hence, either $I(r)$ is negative over $(0,1)$, or it changes sign exactly once, thus verifying the conditions of Lemma 2.30.

As we vary λ and θ (along any curve joining $(0,0)$ to $(1,1)$), we can apply the Implicit Function Theorem to continue the solution, since by the Theorem 2.38, all solutions of (2.169) are non-singular. By the a priori estimates of B. Gidas and J. Spruck [64], the solutions stay bounded, and hence they can be continued for all λ and θ. At $\lambda = 1$ and $\theta = 1$, we conclude existence and uniqueness of positive solutions for our problem. (If there were more than one solution at $\lambda = 1$ and $\theta = 1$, we would have more than one solution at $\lambda = 0$ and $\theta = 0$, a contradiction.) Moreover, the Morse index of this solution is one, since eigenvalues of the linearized problem for (2.169) change continuously, and they cannot cross zero, since solutions are non-singular (and so the number of negative eigenvalues of the linearized problem at $\lambda = 1$ and $\theta = 1$, is the same as at $\lambda = 0$ and $\theta = 0$, i.e., equal to one). $\diamond$

Recently, in [101], we have extended the results of this section to cover the p-Laplacian equations

$$\Delta_p u + f(r,u) = 0, \quad r < 1, \quad u = 0, \quad \text{for } r = 1,$$

including the model case

$$\Delta_p u - a(r)u^{p-1} + b(r)u^q = 0, \quad r < 1, \quad u = 0, \quad \text{for } r = 1.$$

The p-Laplacian case involves considerable technical difficulties. We just mention here the corresponding generalizations of the functions $\xi(r)$ and $\zeta(r)$, which we used above: (where we denote $\varphi(t) = t|t|^{p-2}$)

$$\xi(r) = r^{n-1}\left[(p-1)\varphi(u'(r))w(r) - \varphi'(u'(r))u(r)w'(r)\right] ;$$

$$\zeta(r) = r^n\left[(p-1)\varphi(u'(r))w'(r) + f(r,u(r))w(r)\right] + (n-p)r^{n-1}\varphi(u'(r))w(r).$$

The second one took a while to find (we used *Mathematica* to try out the candidates).

We also mention some recent results on the super-critical problem

$$\Delta u - a(|x|)u + u^p = 0, \quad |x| < 1, \quad u = 0 \text{ for } |x| = 1,$$

with $p > \frac{n+2}{n-2}$, and $a(r)$ positive, see M. Grossi [65], and F. Catrina [31]. It turned out that existence of positive solutions depends on the properties of the Green's function of the linear part, $\Delta u - a(|x|)u$. Observe that for constant $a(r)$, there is no solution, as follows by Pohozaev's identity.

2.12.1 *Radial symmetry for the linearized equation*

Let u be a positive solution of the non-autonomous problem on a unit ball

$$\Delta u + f(|x|, u) = 0, \quad |x| < 1, \quad u = 0 \text{ for } |x| = 1. \tag{2.170}$$

Corresponding linearized problem is

$$\Delta w + f_u(|x|, u)w = 0, \quad |x| < 1, \quad w = 0 \text{ for } |x| = 1. \tag{2.171}$$

Assume that

$$f_r(r, u) \leq 0, \quad \text{for all } r \in [0, 1], \text{ and } u > 0. \tag{2.172}$$

Then by B. Gidas, W.-M. Ni and L. Nirenberg [63], positive solutions of (2.170) are radially symmetric, i.e., $u = u(r)$, with $u'(r) < 0$, and (2.170) becomes

$$u'' + \frac{n-1}{r}u' + f(r, u) = 0, \quad u'(0) = 0, \quad u(1) = 0. \tag{2.173}$$

It turns out that any non-trivial solution of (2.171) is also radially symmetric. This will follow from the following more general result, which is essentially due to C.S. Lin and W.-M. Ni [127], who proved it for the autonomous case, $f = f(u)$.

Theorem 2.39. *([127]) Assume that the condition (2.172) holds, and let* $u = u(r)$ *be a positive solution of (2.170). Let* φ *be an eigenfunction of*

$$\Delta\varphi + f_u(r, u)\varphi + \lambda\varphi = 0, \quad |x| < 1, \quad \varphi = 0 \text{ for } |x| = 1, \quad (2.174)$$

corresponding to an eigenvalue λ. *If* $\lambda \le 0$, *then* φ *must be radially symmetric.*

Proof: Recall that we can decompose

$$\Delta\varphi = \varphi_{rr} + \frac{n-1}{r}\varphi_r + \frac{1}{r^2}\Delta_{S^{n-1}}\varphi,$$

where $\Delta_{S^{n-1}}$ is the Laplace-Beltrami operator on the unit sphere S^{n-1}. Let $0 = \mu_1 < \mu_2 \le \mu_3 \le \cdots$ be the eigenvalues of $-\Delta_{S^{n-1}}$, with $e_1(\theta)$, $e_2(\theta), \ldots$ the corresponding eigenfunctions, $\theta \in S^{n-1}$. It is known that (see e.g., H. Hochstadt [71])

$$\mu_2 = n - 1. \quad (2.175)$$

Set

$$\varphi_k(r) = \int_{S^{n-1}} \varphi(r, \theta)e_k(\theta)\, d\theta.$$

We have

$$\varphi_k''(r) + \frac{n-1}{r}\varphi_k'(r) \quad (2.176)$$
$$= \int_{S^{n-1}} \Delta\varphi\, e_k(\theta)\, d\theta - \frac{1}{r^2}\int_{S^{n-1}} \Delta_{S^{n-1}}\varphi\, e_k(\theta)\, d\theta$$
$$= -f_u(r, u(r))\varphi_k(r) - \lambda\varphi_k(r) + \frac{1}{r^2}\mu_k\varphi_k(r),$$

after an integration by parts, and using (2.174). I.e.,

$$\varphi_k''(r) + \frac{n-1}{r}\varphi_k'(r) + \left[f_u(r, u(r)) - \frac{\mu_k}{r^2}\right]\varphi_k + \lambda\varphi_k = 0. \quad (2.177)$$

Our goal is to show that $\varphi_k \equiv 0$ for all $k \ge 2$. Assume, on the contrary, that $\varphi_k \not\equiv 0$ for some $k \ge 2$. Clearly, $\varphi_k(1) = 0$. Let $0 < \rho_0 \le 1$ be the first zero of $\varphi_k(r)$. We may assume without loss of generality that $\varphi_k(r) > 0$ on $(0, \rho_0)$, and we also have $\varphi_k'(\rho_0) < 0$, by uniqueness of solutions of initial value problems. Differentiating the equation (2.173) with respect to r, gives

$$u_r'' + \frac{n-1}{r}u_r' + \left[f_u(r, u(r)) - \frac{n-1}{r^2}\right]u_r + f_r(r, u(r)) = 0. \quad (2.178)$$

We now multiply the equation (2.177) by u_r, and the equation (2.178) by φ_k, and subtract. Denoting $Q(r) = r^{n-1}(u'\varphi_k' - \varphi_k u'')$, we have

$$Q'(r) + r^{n-1}\left[\frac{n-1}{r^2} - \frac{\mu_k}{r^2}\right]\varphi_k u' + \lambda r^{n-1}\varphi_k u' - r^{n-1}f_r\varphi_k = 0.$$

Using (2.175), it follows that $Q(r)$ is non-increasing on $(0, \rho_0)$. But $Q(0) = 0$, while $Q(\rho_0) = \rho_0^{n-1}u'(\rho_0)\varphi_k'(\rho_0) > 0$, a contradiction. $\diamond$

A different proof of this result can be found in A. Aftalion and F. Pacella [11].

2.13   Exact multiplicity for non-autonomous problems

In P. Korman [93], several exact multiplicity were established for the non-autonomous problem

$$u'' + \frac{n-1}{r}u' + \lambda a(r)g(u) = 0, \ \ u'(0) = 0, \ u(1) = 0, \qquad (2.179)$$

with a positive and non-increasing function $a(r) \in C^1[0,1]$, and $g(u) \in C^2(\bar{R}_+)$, satisfying the following conditions

$$g(0) = 0, \ \text{ and } \ g(u) > 0 \ \text{ for } u \in (0,c), \qquad (2.180)$$

$$g''(u) > 0 \ \text{ when } u \in (0,\alpha), \ \ g''(u) < 0 \ \text{ when } u \in (\alpha,c), \qquad (2.181)$$

for some $0 < \alpha < c \leq \infty$, i.e., $g(u)$ is a convex-concave function. The analysis followed the same steps as for the autonomous problem, however each step was considerably more involved. Even the statements of the results became more technical. In this section, we just review the results, and the details can be found in [93]. As before, a crucial step was to prove positivity of any non-trivial solution of the corresponding linearized problem

$$w'' + \frac{n-1}{r}w' + \lambda a(r)g'(u)w = 0, \ \ w'(0) = 0, \ w(1) = 0. \qquad (2.182)$$

We denote $G(u) = \int_0^u g(t)\,dt$, $\rho = \alpha - \frac{g(\alpha)}{g'(\alpha)}$, and r_1 is the point where $u(r_1) = \rho$.

Theorem 2.40. *Let $u(r)$ be a positive solution of (2.179). Assume that $g(u)$ satisfies the conditions (2.180) and (2.181). Assume that the function $I(r) = 2na(r)G(u(r)) - (n-2)u(r)a(r)g(u(r)) + 2ra'(r)G(u(r))$ satisfies either the condition (i) of Lemma 2.30 with $r_0 = 1$, or the condition (ii) of Lemma 2.30 with $r_0 = r_1$, or the condition (iii) of Lemma 2.30. Assume that*

$$u(r)g(u(r)) - 2G(u(r)) > 0, \ \ \ \ for \ r \in (r_1, 1). \qquad (2.183)$$

Assume that the function $K(u) \equiv \dfrac{ug'(u)}{g(u)}$ satisfies

$$K(u) \ \text{is decreasing on } (0,\rho), \ \ K(\rho) > 1, \qquad (2.184)$$

$$K(u) < K(\rho) \ \text{for } u \in (\rho, c).$$

Assume that the function $a(r)$ is positive, $a'(r) \leq 0$ for $r \in (0,1)$, and the function $\phi(r) \equiv 1 + \frac{1}{2}\frac{ra'(r)}{a(r)}$ is positive and non-increasing on $[0,1)$. Then any non-trivial solution of the linearized problem (2.182) is of one sign, i.e., we may assume that $w(r) > 0$ on $[0,1)$.

To get exact multiplicity results, we shall need one more condition on $a(r)$, which we describe next. Define $b(r) = a(r)^\mu$. We shall assume that

$$(n-1)b(r) + (r^2 b'(r))' + (n-1)rb'(r) > 0, \qquad (2.185)$$
$$\text{for all } r \in [0,1), \text{ and for all } \mu \in [0,1).$$

This condition holds if, say, $a(r)$ is a small perturbation of a positive constant.

A convex-concave $g(u)$ can have the following types of behavior for large u: sublinear, linear, or having a root at some $c > 0$. Correspondingly, we have the following exact multiplicity results. We define $g'(\infty) = \lim_{u \to \infty} \frac{g(u)}{u}$. We denote by $\lambda_1 > 0$ the principal eigenvalue of

$$u'' + \frac{n-1}{r}u' + \lambda a(r)u = 0, \quad u'(0) = 0, \ u(1) = 0. \qquad (2.186)$$

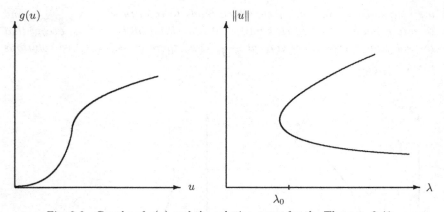

Fig. 2.6   Graphs of $g(u)$ and the solution curve for the Theorem 2.41

Theorem 2.41. *Assume that $a(r)$ and $g(u)$ satisfy the conditions of the Theorem 2.40, and, in addition, the function $a(r)$ satisfies the condition (2.185), and $g'(u) > 0$ for all $u > 0$, with $g'(\infty) = 0$. Assume, first, that $g'(0) = 0$. Then there is a critical $\lambda_0 > 0$, so that for $\lambda < \lambda_0$ the problem (2.179) has no positive solutions, it has exactly one positive solution at $\lambda = \lambda_0$, and exactly two positive solutions for $\lambda > \lambda_0$. Moreover, all solutions lie on a single smooth solution curve, which for $\lambda > \lambda_0$ has two ordered branches $0 < u_-(r, \lambda) < u_+(r, \lambda)$, for all $r \in [0, 1)$ and $\lambda > \lambda_0$, with $u_+(r, \lambda)$ strictly monotone increasing in λ, and $\lim_{\lambda \to \infty} u_+(r, \lambda) = \infty$*

for all $r \in [0, 1)$. For the lower branch we have $\lim_{\lambda \to \infty} u_-(r, \lambda) = 0$ for all $r \in [0, 1)$. In case when $g'(0) > 0$, the situation is similar, except that the lower branch enters zero at $\frac{\lambda_1}{g'(0)}$ (and hence there is exactly one positive solution for $\lambda > \frac{\lambda_1}{g'(0)}$).

The theorem applies, for example, if $g(u) = \frac{u^2}{u^2+1}$.

We consider next the case when $0 < g'(\infty) < \infty$.

Theorem 2.42. *Assume that $a(r)$ and $g(u)$ satisfy the conditions of the Theorem 2.40, and, in addition, the function $a(r)$ satisfies the condition (2.185), and $g'(u) > 0$ for all $u > 0$, with $0 < g'(\infty) < \infty$. Assume also that $\frac{g(u)}{u} < g'(\infty)$ for all $u > 0$. Assume, first, that $g'(0) = 0$. Then for $0 < \lambda < \frac{\lambda_1}{g'(\infty)}$ the problem (2.179) has no positive solutions, and it has a unique positive solution for $\frac{\lambda_1}{g'(\infty)} < \lambda < \infty$. Moreover, all solutions lie on a single smooth solution curve, which tends to infinity when $\lambda \downarrow \frac{\lambda_1}{g'(\infty)}$, and to zero when $\lambda \to \infty$. In case $g'(0) > 0$, the situation is similar, except that the solution curve enters zero at $\frac{\lambda_1}{g'(0)}$ (i.e., there are no positive solutions for $\lambda > \frac{\lambda_1}{g'(0)}$).*

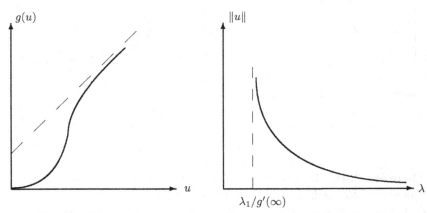

Fig. 2.7   Graphs of $g(u)$ and the solution curve for the Theorem 2.42

The theorem applies, for example, if $g(u) = \frac{u^3}{u^2+1}$.

In the remaining case, when $g(c) = 0$ at some $c < \infty$, we were able to get an exact multiplicity result only in case $n = 1$, see P. Korman [93] for the details.

2.14 Numerical computation of solutions

Good analytical understanding of a problem goes hand in hand with efficient numerical calculation of its solution. We know that for positive solutions the maximum value $u(0) = \alpha$ uniquely determines the solution pair $(\lambda, u(r))$ of the problem

$$u'' + \frac{n-1}{r}u' + \lambda f(u) = 0, \quad r \in (0,1), \quad u'(0) = 0, \ u(1) = 0, \qquad (2.187)$$

see Lemma 2.2 above. We also know that the parameter λ in (2.187) can be "scaled out" i.e., $v(r) \equiv u(\frac{1}{\sqrt{\lambda}}r)$ solves the equation

$$v'' + \frac{n-1}{r}v' + f(v) = 0,$$

while $v(0) = u(0) = \alpha$, and $v'(0) = u'(0) = 0$. The first root of $v(r)$ is $r = \sqrt{\lambda}$. We, therefore, solve the initial value problem

$$v'' + \frac{n-1}{r}v' + f(v) = 0, \quad v(0) = \alpha, \ v'(0) = 0, \qquad (2.188)$$

and find its first positive root r. Then $\lambda = r^2$, by the above remarks. This way for each α we can find the corresponding λ. After we choose sufficiently many α_n's and compute the corresponding λ_n's, we plot the pairs (λ_n, α_n), obtaining a bifurcation diagram in (λ, α) plane. We stress that the resulting two-dimensional bifurcation curve gives a faithful representation of the solution set of (2.187), since the value $u(0) = \alpha$ uniquely determines the solution pair $(\lambda, u(r))$. (If one needs to compute the actual solution $u(r)$ of (2.187) at the point $(\lambda, u(0) = \alpha)$, this can be easily done by using the **NDSolve** command in *Mathematica*.) This program for producing a bifurcation diagram of (2.187), consists of essentially one short loop. It can be found at the author's web-page: http://math.uc.edu/~kormanp/. We used this program to produce all of the bifurcation diagrams in this chapter.

We see absolutely no need to ever use finite differences (or finite elements) for the problem (2.187). To explain why, it suffices to consider a one-dimensional version of our problem (2.187):

$$u''(x) + \lambda f(u(x)) = 0 \quad \text{for } -1 < x < 1, \ u(-1) = u(1) = 0. \qquad (2.189)$$

If we divide the interval $(-1, 1)$ into n pieces, with step $h = 2/n$, and subdivision points $x_i = ih$, and denote by u_i the numerical approximation of $u(x_i)$, the finite difference approximation of (2.189) is

$$\frac{u_{i+1} - 2u_i + u_{i-1}}{h^2} + \lambda f(u_i) = 0, \quad 1 \le i \le n-1, \ u_0 = u_n = 0. \qquad (2.190)$$

This is a system of *nonlinear* algebraic equations, more complicated in every way than the original problem (2.189). In particular, this system often has more solutions than the corresponding differential equation (2.189). The existence of the extra solutions of difference equations (not corresponding to the solutions of the differential equation (2.189)) has been recognized for a while, and a term *spurious solutions* has been used, see e.g., K. Schmitt, R.C. Thompson and W. Walter [165]. For example, in case $f(u) = e^u$, the solution curve of (2.189) has exactly one turn (as we saw before), while the solution curve of (2.190) has three turns, see P. Korman [89]. Increasing the number of subdivision points n, does not remove the two spurious turns, it just moves them closer to the line $\lambda = 0$. (Actually, the spurious turns are avoided in the opposite direction, when $n \leq 6$, see [89]. We found this hard to prove, even when $n = 2$.) When studying the algebraic system (2.190), we can no longer rely on the familiar tools from differential equations. Even in the simplest case of $f(u) = u^k$, the analysis of the problem (2.190) is very involved, see E.L. Allgower [7]. We avoid spurious solutions, with the approach described at the beginning of this section.

2.14.1   *Using power series approximation*

As we had just mentioned, solving the initial value problem

$$u'' + \frac{n-1}{r}u' + f(u) = 0 \tag{2.191}$$

$$u(0) = u_0, \quad u'(0) = 0,$$

is easily done using standard methods, or just using **NDSolve** command in *Mathematica*. The singularity ar $r = 0$ is easily circumvented, for example, by choosing $\epsilon > 0$ small, we express $u(\epsilon) \simeq u_0 + \frac{1}{2}u''(0)\epsilon^2 = u_0 - \frac{1}{2n}f(u_0)\epsilon^2$, $u'(\epsilon) \simeq u''(0)\epsilon = -\frac{1}{n}f(u_0)\epsilon$, and then solve the resulting non-singular initial value problem for $r > \epsilon$. For that, one can use either the classical Runge-Kutta method, or way more sophisticated procedures, which programs like *Mathematica* employ. Next, we discuss the use of power series as an alternative numerical method, thus diverging somewhat from the theme of this book. This subsection, and the next one, are based on P. Korman [95].

The idea of Runge-Kutta method is to replace a five term Taylor approximation with function evaluations. This method was developed way before the advent of computers. Since symbolic capabilities have become widely available recently, one can easily perform repeated differentiations of $f(u)$, needed for Taylor approximations. A major attraction of power series method is that one can easily estimate the truncation error, making possible *exact* computations (i.e., numerical computations with an error bound available at each step). We begin by showing that the power series solution contains only the even powers of r, i.e., $u(r) = \Sigma_{i=0}^{\infty} a_i \, r^{2i}$, with $a_0 = u_0$. We then derive a recursive formula for a_i, which is easy to implement in *Mathematica*. One can quickly compute a lot of terms of the series, giving highly accurate approximation for small r, practically an exact solution. The problem is that the power series solution usually has a finite radius of convergence, which can be small (if $f(u_0)$ is large). We need another approximation of solutions, which is defined for all r, and which agrees with the power series approximation for small r. A classical way of doing so is to use Pade approximations. These are rational functions, whose power series expansion agrees, up to a certain degree, with that of the function being approximated. We shall see the effectiveness of Pade approximations below.

Lemma 2.31. *Any solution of the problem (2.191) is an even function.*

Proof: Even though the initial value problem has a singularity at $r = 0$, it has the properties of regular initial value problems, in particular there is existence and uniqueness of solutions to (2.191) for small $|r|$, as follows by Lemma 2.1. We may consider solutions of (2.191) in case $r < 0$ too. Observe that the change of variables, $r \to -r$, leaves (2.191) invariant. If solution was not even, we would have two different solutions of (2.191) for $r > 0$, $u(r)$ and $u(-r)$, contradicting the just mentioned uniqueness result.

$\diamondsuit$

In view of this lemma, we may express the solution of (2.191) as $u(r) = \Sigma_{i=0}^{\infty} a_i \, r^{2i}$. Numerically, we shall of course be computing the partial sums $u(r) = \Sigma_{i=0}^{k} a_i \, r^{2i}$, which we expect to provide us with a solution, up to the terms of order $O(r^{2k+2})$. Of course, $a_i = \frac{u^{(2i)}(0)}{(2i)!}$.

Theorem 2.43. *([95]) Consider the problem (2.191), with $f(u) \in C^{\infty}(R)$. Then the coefficients of the Taylor series of its solution $u(r)$ are computed inductively for $k = 1, 2, \ldots$*

$$a_k = -\frac{1}{2k \, (n + 2k - 2) \, (2k - 2)!} \, \frac{d^{2k-2}}{dr^{2k-2}} f(\bar{u}(r))|_{r=0} \,, \qquad (2.192)$$

where $\bar{u}(r) = \Sigma_{i=0}^{k-1} a_i \, r^{2i}$ involves the previously computed a_i's, and $a_0 = u_0$.

Proof: Beginning with $\bar{u}(r) = u_0$, we write inductively

$$u(r) = \bar{u}(r) + a_k r^{2k}, \quad k = 1, 2, \ldots,$$

where we regard $\bar{u}(r)$ as already computed, and the question is how to get a_k. If we plug this ansatz into (2.191), we have

$$2k(n + 2k - 2) a_k r^{2k-2} = -f\left(\bar{u}(r) + a_k r^{2k}\right) - \bar{u}'' - \frac{n-1}{r}\bar{u}' + O(r^{2k+2}),$$

and on the surface it may appear that we need to solve a nonlinear equation for a_k. It turns out that we can get an explicit formula for a_k. Express

$$a_k = -\frac{1}{\gamma_k} \lim_{r \to 0} \frac{f\left(\bar{u}(r) + a_k r^{2k}\right) + \bar{u}'' + \frac{n-1}{r}\bar{u}'}{r^{2k-2}}, \qquad (2.193)$$

where $\gamma_k = 2k(n + 2k - 2)$. Using two term Taylor expansion of the first function, we have

$$\lim_{r \to 0} \frac{f\left(\bar{u}(r) + a_k r^{2k}\right) + \bar{u}'' + \frac{n-1}{r}\bar{u}'}{r^{2k-2}} = \lim_{r \to 0} \frac{f\left(\bar{u}(r)\right) + \bar{u}'' + \frac{n-1}{r}\bar{u}'}{r^{2k-2}}.$$

Observe that when we repeatedly apply the L'Hôspital's rule $2k - 2$ times, all terms involved in $\bar{u}'' + \frac{n-1}{r}\bar{u}'$ will vanish (since r^{2k-4} is the highest power there), and we have

$$\lim_{r \to 0} \frac{f\left(\bar{u}(r)\right) + \bar{u}'' + \frac{n-1}{r}\bar{u}'}{r^{2k-2}} = \frac{1}{(2k-2)!} \frac{d^{2k-2}}{dr^{2k-2}} f(\bar{u}(r))\big|_{r=0}.$$

Using this in (2.193), we conclude the proof.   ◇

Example We have solved the problem

$$u'' + \frac{2}{r}u' + u^5 = 0, \quad u(0) = 1, \ u'(0) = 0,$$

which gives, in the dimension $n = 3$, a radial solution of the critical PDE, with applications to geometry,

$$\Delta u + u^{\frac{n+2}{n-2}} = 0,$$

where n is the dimension of the space. Using the formula (2.192), we had computed a series solution, which begins with

$$u(r) = 1 - \frac{1}{6}r^2 + \frac{1}{24}r^4 - \frac{5}{432}r^6 + \frac{35}{10368}r^8 - \frac{7}{6912}r^{10} + \cdots. \qquad (2.194)$$

There is a closed form solution for the critical initial value problem

$$u'' + \frac{n-1}{r}u' + u^{\frac{n+2}{n-2}} = 0, \quad u(0) = \alpha, \ u'(0) = 0.$$

It is

$$u(r) = \alpha \left(\frac{n(n-2)}{n(n-2) + \alpha^{4/(n-2)} r^2}\right)^{\frac{n-2}{2}},$$

discovered independently by T. Aubin [16] and G. Talenti [179] in their study of the *best Sobolev imbedding constants*. It is easy to check that the series (2.194) corresponds to this solution, when $\alpha = 1$, $n = 3$.

Example On a unit ball in R^3 we consider a sub-critical problem

$$\Delta v + v^4 = 0, \quad \text{for } |x| < 1, \ v = 0 \text{ when } |x| = 1. \tag{2.195}$$

By the classical results of B. Gidas, W.-M. Ni and L. Nirenberg [63], the problem (2.195) has a unique positive solution, which is moreover radially symmetric, i.e., $u = u(r)$, where $r = |x|$. It follows that (2.195) may be replaced by the ODE

$$v'' + \frac{2}{r}v' + v^4 = 0, \quad v'(0) = 0, \ v(1) = 0. \tag{2.196}$$

We begin by solving the initial-value problem

$$u'' + \frac{2}{r}u' + u^4 = 0, \quad u(0) = 1, \ u'(0) = 0.$$

After computing 55 terms of this series (powers of up to r^{110}), we have used a built in *Mathematica* command to approximate this polynomial of power 110 by its Pade approximation, centered at $r = 0$, and using polynomials of power 50 for both the numerator and denominator of Pade approximation. We used this approximation of $u(r)$ to find its first root $R \approx 14.9716$. Then by standard scaling, $v(r) = R^{2/3}u(Rr)$ gives an approximation to the solution of (2.196). To see how good is this approximation, we have computed the left hand side of (2.196)

$$q(r) = v'' + \frac{2}{r}v' + v^4.$$

In Figure 2.8, we give the graph of $q(r)$, which shows that the accuracy is high. ($q(r)$ is even smaller when $r \in (0, 0.6)$.) In Figure 2.9, we graph the approximation of $v(r)$ that we have obtained ($v(r) = R^{2/3}u(Rr)$).

2.14.2 *Application to singular solutions*

We consider positive solutions of the Dirichlet problem on a unit ball in R^n

$$\Delta u + \lambda u + u^p = 0, \quad \text{for } |x| < 1, \ u = 0 \text{ on } |x| = 1. \tag{2.197}$$

Here λ is a positive parameter, and $p > 1$. It is well known that when $p > \frac{n+2}{n-2}$, the critical exponent, there is a unique $\bar{\lambda}$ at which a singular solution exists (i.e., $u(r)$ is unbounded at $r = 0$), see F. Merle and L. A. Peletier [134]. Moreover, the significance of the singular solution is by now

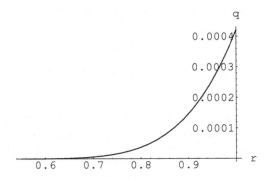

Fig. 2.8   Defect of the approximation of the solution

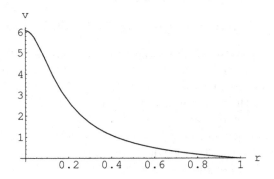

Fig. 2.9   Solution of the Dirichlet problem (2.196)

well understood: the curve of positive solutions of (2.197) tends to infinity, as $\lambda \to \bar{\lambda}$, and it approaches the singular solution for $r \neq 0$. Using power series expansion, we now give a formal derivation that there is a unique $\bar{\lambda}$ at which a singular solution exists for $p > \frac{n}{n-2}$, i.e., for some sub-critical equations too.

It is easy to see that a singular solution must blow up like $r^{-\frac{2}{p-1}}$ near $r = 0$. We denote $\theta = -\frac{2}{p-1}$, and $A = -\theta(n + \theta - 2)$. Observe that $A > 0$ if and only if $p > \frac{n}{n-2}$. We shall obtain a solution of (2.197) in the form

$$u(r) = r^{-\frac{2}{p-1}} v(r), \tag{2.198}$$

where $v(r)$ satisfies

$$v(0) = A^{\frac{1}{p-1}}. \tag{2.199}$$

We assume that $n \geq 3$, and $p > \frac{n}{n-2}$. In case $p < \frac{n+4}{n}$, we assume additionally that (this assumption is not needed if $p \geq \frac{n+4}{n}$)

$$2i\left(2i + n + 2\theta - 2\right) + (p-1)A > 0 \quad \text{for all integers } i \geq 1. \tag{2.200}$$

Observe that (2.200) gives a finite set of conditions, since the condition $p > \frac{n}{n-2}$ implies that $\theta > -n + 2$, and then (2.200) holds if $i > \frac{n-2}{2}$. If we assume a stronger condition $p \geq \frac{n+4}{n}$ ($\frac{n+4}{n} > \frac{n-2}{2}$, for $n \geq 4$), then $\theta > -n/2$, and the condition (2.200) holds. Since $\frac{n+4}{n} < \frac{n+2}{n-2}$, we see that the condition (2.200) is satisfied for supercritical p.

Radial solutions of (2.197) satisfy

$$u'' + \frac{n-1}{r}u' + \lambda u + u^p = 0, \quad \text{for } r < 1, \quad u(1) = 0. \tag{2.201}$$

Using the ansatz (2.198) in (2.201), we see that $v(r)$ satisfies

$$r^2 v'' + (n - 1 + 2\theta)rv' + \left[\theta(n + \theta - 2) + \lambda r^2\right]v + v^p = 0. \tag{2.202}$$

We are looking for a solution of (2.202) satisfying

$$v'(0) = v(1) = 0. \tag{2.203}$$

We now scale λ out of (2.202), by letting $r = \frac{1}{\sqrt{\lambda}}\xi$. Then $v(\xi)$ satisfies

$$\xi^2 v'' + (n - 1 + 2\theta)\xi v' + \left[-A + \xi^2\right]v + v^p = 0, \tag{2.204}$$

and $\frac{dv}{d\xi}(0) = 0$. Setting here $\xi = 0$, we see that $v(0)$ satisfies the condition (2.199). Hence, the condition (2.199) is necessary for existence of singular solution. We shall produce a power series solution of (2.204).

The equation (2.204) can be considered for $\xi < 0$ too. Since this equation is invariant under the transformation $\xi \to -\xi$, it follows that any classical solution is an even function, and so its power series expansion contains only the even powers. We then look for solution of (2.204) in the form

$$v(\xi) = \Sigma_{i=0}^{\infty} a_i \xi^{2i}. \tag{2.205}$$

Plugging this into (2.204), we have

$$a_0 + \Sigma_{i=1}^{\infty} c_i \xi^{2i} + \left(a_0 + \Sigma_{i=1}^{\infty} a_i \xi^{2i}\right)^p = 0, \tag{2.206}$$

where we denote

$$c_i = \left[2i(2i + n + 2\theta - 2) - A\right]a_i + a_{i-1}. \tag{2.207}$$

The constant terms in (2.206) will vanish, if we choose $a_0 = A^{\frac{1}{p-1}}$. We show next that setting the power of ξ^{2i} to zero, produces a linear equation for a_i of the form

$$[2i(2i + n + 2\theta - 2) + (p-1)A]\, a_i + g\,(a_0, a_1, \ldots, a_{i-1}) = 0, \qquad (2.208)$$

where g is some function of previously computed coefficients. (Actual computations are easy to program, using *Mathematica*.) Indeed, denoting $t = \Sigma_{i=1}^{\infty} a_i \xi^{2i}$, we rewrite the last term in (2.206) as

$$\left(a_0 + \Sigma_{i=1}^{\infty} a_i \xi^{2i}\right)^p = a_0^p\, \Sigma_{j=0}^{p} \begin{pmatrix} p \\ j \end{pmatrix} \frac{t^j}{a_0^j}.$$

The term ξ^{2i}, with a coefficient involving a_i occurs, when $j = 1$ in this sum, justifying (2.208). Using (2.208), we compute all a_i's. Let γ denote the first root of $v(\xi)$. Then $\bar{\lambda} = \gamma^2$, and $u(r) = r^\theta u(\gamma r)$ is the singular solution of (2.201) at $\lambda = \bar{\lambda}$.

Our numerical computations strongly suggest convergence of the series (2.205), but we did not carry out the proof.

Example If $n = 3$, then $\frac{n+2}{n-2} = 5$, $\frac{n}{n-2} = 3$, and $\frac{n+4}{n} = 7/3$. We took a subcritical $p = 4$, and found that the singular solution of

$$u'' + \frac{n-1}{r}u' + \lambda u + u^4 = 0, \ \text{ for } r < 1, \ \ u(1) = 0 \qquad (2.209)$$

occurs at $\bar{\lambda} \simeq 5.26295$. Here are the first few terms of the singular solution

$$u(r) \simeq r^{-2/3}(0.60571 - 0.79695r^2 + 0.18236r^4 + 0.02615r^6 - 0.02201r^8$$
$$+\, 0.00471r^{10} + 0.00054r^{12}).$$

Its graph is shown in Figure 2.10. When we plugged this approximation of the solution into (2.209), at $\lambda = \bar{\lambda}$, we obtained the left hand side practically zero for all $r \in (0, 1)$.

2.15   Radial solutions of Neumann problem

We consider Neumann problem on a unit ball $B \subset R^n$

$$\Delta u(x) + \lambda f(u(x)) = 0, \ \ x \in B, \ \ \frac{\partial u}{\partial \nu} = 0 \ \text{ for } x \in \partial B, \qquad (2.210)$$

where ν is the unit normal vector, pointing outside of B, and λ is a positive parameter. This problem has been much less studied than the corresponding Dirichlet problem, perhaps because there is no analog of B. Gidas,

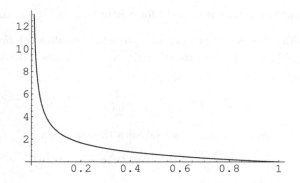

Fig. 2.10   Singular solution of the problem (2.209)

W.-M. Ni and L. Nirenberg theorem [63]. However, it is still reasonable to study radial and decreasing solutions, which we will do in this section, based on P. Korman [98]. We shall assume that $f(u)$ satisfies

$$f(u) \in C^1[0, \infty), \quad f(0) = 0, \text{ and for some } b > 0 \text{ we have:} \qquad (2.211)$$

$$f(u) < 0 \text{ on } (0, b), \text{ and } f(u) > 0 \text{ for } b < u < c, \text{ with } c \leq \infty;$$

$$f'(0) < 0, \quad f'(b) > 0; \quad f(c) = 0, \text{ in case } c < \infty.$$

For example, $f(u) = -u + u^p$, $p > 1$, or $f(u) = u(u - b)(c - u)$. Under these assumptions, it is natural to consider positive solutions, i.e., we consider the problem

$$u''(r) + \frac{n-1}{r} u'(r) + \lambda f(u) = 0, \quad 0 < r < 1, \qquad (2.212)$$

$$u'(0) = u'(1) = 0,$$

$$u(r) > 0, \quad u'(r) < 0, \quad 0 < r < 1.$$

We shall study solution curves for the Neumann problem (2.212), with $f(u)$ satisfying (2.211). Since solutions of (2.212) are decreasing, and $u''(0) = -\frac{\lambda}{n} f(u(0))$, it follows that $u(0) > b$. Similarly, $u''(1) = -\lambda f(u(1))$, which implies that $u(1) < b$ (the possibilities of $u(0) = b$, or of $u(1) = b$, are ruled out by the uniqueness theorem for initial value problems, which would imply $u \equiv b$).

The linearized problem, corresponding to (2.212), is

$$w''(r) + \frac{n-1}{r} w'(r) + \lambda f'(u)w = 0, \quad 0 < r < 1, \qquad (2.213)$$

$$w'(0) = w'(1) = 0.$$

The following lemma will be used for continuation of Neumann branches.

Lemma 2.32. *Assume that $u(r)$ is a singular solution of the Neumann problem (2.212), i.e., the linearized problem (2.213) has a non-trivial solution $w(r)$. Assume $f(u) \in C^1(R)$. Then*

$$\int_0^1 f(u)wr^{n-1}\, dr = -\frac{1}{2\lambda}w(1)u''(1). \qquad (2.214)$$

Proof: The function $z(r) = ru_r$ satisfies the equation

$$z''(r) + \frac{n-1}{r}z'(r) + \lambda f'(u)z = -2\lambda f(u). \qquad (2.215)$$

Combining this with (2.213),

$$\left[r^{n-1}\left(z'w - zw'\right)\right]' = -2\lambda f(u)wr^{n-1}.$$

Integrating over $(0,1)$, we obtain (2.214). $\diamond$

Remark Observe that we did not need the assumptions (2.211) on $f(u)$. Also, this lemma holds for any solution of Neumann problem, not necessary decreasing or positive.

We may assume that any non-trivial solution of (2.213) satisfies $w(0) > 0$. We have the following rather general lemma.

Lemma 2.33. *Let $u(r)$ be a solution of (2.212), and assume that $f(u)$ satisfies (2.211). Then any non-trivial solution of (2.213) must be sign changing.*

Proof: Assume, on the contrary, that $w(r) > 0$. Differentiating the equation (2.212), we have

$$u''' + \frac{n-1}{r}u''(r) + \lambda f'(u)u' - \frac{n-1}{r^2}u' = 0.$$

Combining this with (2.213),

$$\left[r^{n-1}\left(w'u' - wu''\right)\right]' + (n-1)r^{n-3}u'w = 0.$$

It follows that the function $q(r) \equiv r^{n-1}\left(w'u' - wu''\right)$ is increasing on $(0,1)$. But $q(0) = 0$, while

$$q(1) = -w(1)u''(1) = \lambda w(1)f(u(1)) < 0,$$

which is a contradiction. $\diamond$

We were able to prove that solutions of the Neumann problem are non-singular, under the condition

$$\frac{f(u)}{u-b} > f'(u) \quad \text{for all } u > 0. \qquad (2.216)$$

This condition (going back to Opial [141] in the ODE case, see also R. Schaaf [161], and P. Korman [92]) is rather restrictive. It holds if e.g., $f''(u)(u - b) < 0$ for $u > 0$, i.e., $f''(u) < 0$ (> 0) for $u > b$ ($< b$).

Lemma 2.34. *Assume that $f(u)$ satisfies the conditions (2.211) and (2.216). Then the problem (2.213) has no non-trivial solutions.*

Proof: Let $\xi \in (0, 1)$ be the point where $u(\xi) = b$. According to the preceding lemma, $w(r)$ has to change sign on $(0, 1)$. However, we shall show that $w(r)$ cannot vanish on $(0, \xi]$, and on $[\xi, 1)$.

We rewrite the equation in (2.212) in the form

$$(u - b)''(r) + \frac{n - 1}{r} (u - b)'(r) + \lambda \frac{f(u)}{u - b} (u - b) = 0. \qquad (2.217)$$

Multiplying (2.217) by w, the equation (2.213) by $u - b$, and subtracting

$$\left[r^{n-1}(wu' - w'(u - b))\right]' + \lambda r^{n-1} \left[\frac{f(u)}{u - b} - f'(u)\right](u - b)w = 0. \qquad (2.218)$$

If we now assume that $w(r)$ vanishes on $(0, \xi]$, we can find $\eta \in (0, \xi]$ (the first zero), so that $w(r) > 0$ on $(0, \eta)$, and $w(\eta) = 0$. Integrating (2.218) over $(0, \eta)$, we obtain

$$-\eta^{n-1}w'(\eta)(u(\eta) - b) + \lambda \int_0^{\eta} r^{n-1} \left[\frac{f(u)}{u - b} - f'(u)\right](u - b)w \, dr = 0.$$

Since both quantities on the left are positive, we have a contradiction. In case $w(r)$ vanishes on $[\xi, 1)$, we can find $\eta \in [\xi, 1)$ (the last zero), so that $w(r) > 0$ on $(\eta, 1]$, and $w(\eta) = 0$. Integrating (2.218) over $(\eta, 1)$, we have

$$\eta^{n-1}w'(\eta)(u(\eta) - b) + \lambda \int_{\eta}^{1} r^{n-1} \left[\frac{f(u)}{u - b} - f'(u)\right](u - b)w \, dr = 0.$$

Since $u(r) < b$ over $(\eta, 1]$, both terms on the left are negative, which is a contradiction. $\qquad \diamond$

The following lemma is similar to what we had in the Dirichlet case.

Lemma 2.35. *The value of $u(0) = \alpha$ uniquely identifies the solution pair $(\lambda, u(r))$ of the Neumann problem (2.212) (i.e., there is at most one λ, with at most one solution $u(r)$, so that $u(0) = \alpha$).*

Proof: Assume, on the contrary, that we have two solution pairs $(\lambda, u(r))$ and $(\mu, v(r))$, with $u(0) = v(0) = \alpha$. Clearly, $\lambda \neq \mu$, since otherwise we have a contradiction with uniqueness of initial value problems, see Lemma

2.1. Then $u(\frac{1}{\sqrt{\lambda}}r)$ and $v(\frac{1}{\sqrt{\mu}}r)$ are both solutions of the same initial value problem

$$u''(r) + \frac{n-1}{r}u'(r) + f(u) = 0 \quad u(0) = \alpha, \ u'(0) = 0,$$

and hence $u(\frac{1}{\sqrt{\lambda}}r) = v(\frac{1}{\sqrt{\mu}}r)$, but that is impossible, since the derivative of the first function has its first root at $r = \sqrt{\lambda}$, while the derivative of the second one has its first root at $r = \sqrt{\mu}$.   ◇

Before we continue with Neumann problem, let us consider "shooting", i.e., we solve the initial value problem

$$u'' + \frac{n-1}{r}u' + f(u) = 0, \ u(0) = \alpha, \ u'(0) = 0, \qquad (2.219)$$

for various values of $\alpha > 0$. We shall consider only those α's for which $f(\alpha) > 0$, and consequently $u''(0) = -\frac{1}{n}f(\alpha) < 0$. It follows that for small $r > 0$, the function $u(r)$ is positive and decreasing:

$$u(r) > 0 \ \text{ and } \ u'(r) < 0. \qquad (2.220)$$

We continue the solution $u(r)$ for increasing r, while (2.220) holds. I.e., we stop as soon as (2.220) is violated. This may happen in two ways: solution becomes zero, or it develops zero slope. If $u(R) = 0$, then we make a change of variables $r = R\xi$, and see that the function $u(\xi)$ satisfies the Dirichlet problem

$$u'' + \frac{n-1}{r}u' + \lambda f(u) = 0, \quad u'(0) = 0, \ u(1) = 0, \qquad (2.221)$$

with $\lambda = R^2$. We refer to the solution curves of (2.221) as *Dirichlet curves* (similarly solutions of (2.212) are *Neumann curves*). Clearly, $u|_{\xi=0} = \alpha$. We then say that the initial value α belongs to *Dirichlet range*. Similarly, if $u'(R) = 0$, then using the same change of variables $r = R\xi$, we see $u(\xi)$ satisfies the Neumann problem (2.212), with $\lambda = R^2$. Then α belongs to *Neumann range*. Conversely, all values of $\alpha = u(0)$ assumed by any solution of the Dirichlet problem (2.221) belong to the Dirichlet range (with similar correspondence of Neumann curve and Neumann range). So, $\alpha = u(0)$ belongs to Dirichlet (Neumann) range if and only if for some λ there is a solution of (2.221) (of (2.212)), with $u(0) = \alpha$.

The following theorem describes the Neumann curve bifurcating from $u \equiv b$ at $\lambda = \lambda_1/f'(b)$, where $\lambda_1 > 0$ is the first positive eigenvalue of $-\Delta$ on a unit ball, with zero Neumann condition.

Theorem 2.44. *Consider the Neumann problem (2.212), with $f(u)$ satisfying (2.211). Assume that the Dirichlet range is non-empty for the problem (2.219). Then there is a curve of solutions of (2.212), bifurcating from the trivial solution $u = b$, at $\lambda = \lambda_1/f'(b)$. This curve continues globally, with the maximum value $u(0)$ strictly increasing, and with λ eventually tending to infinity. Moreover, $\lim_{\lambda \to \infty} u(0, \lambda) > \theta$, where we define θ by $\int_0^\theta f(u)\,du = 0$. Any other solution curve of (2.212) makes at least one turn to the right, with both branches extending to infinity along the λ axis (i.e., λ tends to infinity on both branches).*

If, in addition, the linearized problem (2.213) admits only the trivial solution (this happens if e.g., the condition (2.216) holds), then the solution curve bifurcating from $u = b$ does not turn, and there are no other solution curves of (2.212).

Proof: According to the Crandall-Rabinowitz theorem on bifurcation from simple eigenvalue [39], there is a curve of solutions of (2.212), bifurcating from the trivial solution $u = b$, at $\lambda = \lambda_1/f'(b)$ (on a ball, all radial Neumann eigenvalues are simple, since the value of $w(1)$ parameterizes the eigenfunctions). Solutions on the bifurcating curve are positive (since they are close to b), and decreasing (since $u - b$ is asymptotically proportional to the first eigenfunction, as $\lambda \to \lambda_1$). For the same reason, the maximum value $u(0, \lambda)$ is increasing along the solution curve. We claim that all three of these properties are preserved along the solution curve. Indeed, the maximum value remains increasing, in view of Lemma 2.35. Writing the equation (2.212) as $\left(r^{n-1}u'\right)' = -\lambda r^{n-1}f(u)$, we see that $r^{n-1}u' < 0$ on $(0, 1)$ (since $f(u(r))$ changes sign exactly once, the function $r^{n-1}u'$ is first decreasing, then increasing, and it is zero at the end-points), and so $u(r)$ is decreasing. Finally, if $u(r, \lambda)$ were to become zero, this would have to happen at $r = 1$ (since $u(r)$ is decreasing), where we also have $u'(1) = 0$. Since $f(0) = 0$, we would then have $u \equiv 0$, by uniqueness for initial value problems, contradicting $u(0, \lambda) > b$.

We now continue this curve. If the linearized problem (2.213) admits only the trivial solution, we may use the Implicit Function Theorem to continue this solution curve in λ. We show next that at singular solutions of (2.212), i.e., when the linearized problem (2.213) admits non-trivial solutions, the Crandall-Rabinowitz Theorem 1.2 applies. If $w(r)$ is a non-trivial solution of (2.213), then $w(1) \neq 0$ (since otherwise $w(r) \equiv 0$). As we mentioned above, for decreasing solutions of (2.212) $u(0) > b$, while $u(1) < b$.

This implies that $u''(1) = -\lambda f(u(1)) > 0$, and hence by (2.214),

$$\int_0^1 f(u(r))\, w(r) r^{n-1}\, dr = -\frac{1}{2\lambda} w(1) u''(1) \neq 0. \qquad (2.222)$$

We now recast the problem (2.212) in the operator form $F(\lambda, u) : R \times \{H^2(0,1) \,|\, u'(0) = u'(1) = 0\} \to L^2(0,1)$, with

$$F(\lambda, u) = u''(r) + \frac{n-1}{r}\, u'(r) + \lambda f(u).$$

The linearized operator $F_u(\lambda, u)w$ is given by the left hand side of the equation (2.213), and hence its null space is one dimensional, spanned by $w(r)$. Since $F(\lambda, u)$ is a Fredholm operator, its range has codimension one. And the crucial condition, $F_\lambda \notin R(F_u)$, holds in view of (2.222). Hence, the Crandall-Rabinowitz Theorem 1.2 applies, and we have a curve of H^2 solutions through a critical point, which we then boot-strap to classical solutions.

It follows that any Neumann curve extends globally. Next, we observe that there is a "free" a priori estimate for Neumann curves: the maximum value of solution lies below that of any solution of the corresponding Dirichlet problem. Indeed, since the Dirichlet range is non-empty, the Dirichlet problem (2.221) has solution at some λ. This solution lies on a solution curve, which extends to the upper limit $c \leq \infty$. Hence, all sufficiently large values of $u(0)$ belong to the Dirichlet range, and so Neumann curves cannot go there. Hence, Neumann solution curves cannot go to infinity at a finite λ. It follows that the curve, bifurcating from $u \equiv b$, has one end where $\lambda \to \infty$, while for any other curve both ends extend to infinity in λ, which implies existence of at least one turn to the right on such curves. $\Diamond$

2.15.1   *A computer assisted study of ground state solutions*

We shall now try to understand all solutions of the initial value problem (2.219), regardless of their boundary behavior. The value of $\alpha = u(0)$ may belong to either Dirichlet or Neumann range, or it may give rise to a *ground state*, i.e., solution of

$$u''(r) + \frac{n-1}{r} u'(r) + f(u) = 0, \quad 0 < r < \infty, \qquad (2.223)$$

$$u'(0) = u(\infty) = 0,$$

$$u(r) > 0, \ u'(r) < 0, \quad 0 < r < \infty.$$

Let us assume that $f(u)$ satisfies the conditions in (2.211), with $c = \infty$.

We begin by observing that the Dirichlet range might be empty, as the following example from P. Korman and Y. Li [107] shows.

Example Consider a case of super-critical $f(u)$

$$u'' + \frac{n-1}{r}u' - u^{\frac{n+2}{n-2}} + u^p = 0, \quad u(0) = \alpha, \quad u'(0) = 0, \qquad (2.224)$$

with $p > \frac{n+2}{n-2}$. Our nonlinearity $f(u) = -u^{\frac{n+2}{n-2}} + u^p$ satisfies the condition (2.211). We claim that for any $\alpha > 0$, $u(r) > 0$ for all r, which means that the Dirichlet range is empty. Recall the well-known Pohozhaev's identity, which we used before, and which is easy to check, using the equation in (2.223). It says that, with $H(r) \equiv \frac{1}{2}\left[ru'^2 + (n-2)uu'\right] + rF(u)$, and $F(u) = \int_0^u f(t)\,dt$, we have

$$H' + \frac{n-1}{r}H = nF - \frac{n-2}{2}u\,f(u) = u^{p+1}\left(\frac{n}{p+1} - \frac{n-2}{2}\right) < 0\,,$$

which is the same as

$$\left[r^{n-1}H(r)\right]' < 0\,. \qquad (2.225)$$

Assume, on the contrary, that $u(r)$ vanishes at some $r = \xi$. Integrating (2.225) over $(0, \xi)$, we get

$$0 \le \frac{1}{2}\xi^n u'^2(\xi) < 0\,,$$

a contradiction.

We shall, therefore, assume that $f(u)$ is sub-critical, and also that it is asymptotic to a power, i.e., for some $1 < p < \frac{n+2}{n-2}$, and $a > 0$, we have

$$\lim_{u \to \infty} \frac{f(u)}{u^p} = a\,. \qquad (2.226)$$

The following lemma is known, so we just sketch its proof.

Lemma 2.36. *Assume that $f(u)$ satisfies the conditions (2.211) and (2.226). Then all α's sufficiently large belong to the Dirichlet range, i.e., the corresponding solution of (2.219) is decreasing, and it becomes zero at some $r = r_0$.*

Proof: Write $f(u) = g(u) + au^p$, with $\lim_{u \to \infty} \frac{g(u)}{u^p} = 0$. In (2.219) we set $u = \alpha w$, $r = \frac{1}{\alpha^\beta}\xi$. Letting $2\beta = p - 1$, and using primes for the derivatives of $w(\xi)$, we get

$$w'' + \frac{n-1}{\xi}w' + \frac{g(\alpha w)}{\alpha^p} + aw^p = 0, \quad w(0) = 1, \quad w'(0) = 0\,. \qquad (2.227)$$

For the problem

$$w'' + \frac{n-1}{\xi} w' + aw^p = 0, \quad w(0) = 1, \quad w'(0) = 0,$$

it is well known that the solution is decreasing, becoming zero at some $r = R$, and then it continues to decrease, taking negative values. By continuity, for α large, the solution of (2.227) is decreasing, becoming zero near $r = R$, since the term $\frac{g(\alpha w)}{\alpha^p}$ is uniformly small for $r \in (0, R)$, when α is large. $\diamond$

We know that the Dirichlet range extends to infinity. If there is only one Neumann curve, then the Neumann range is connected, and everything in between the Dirichlet and Neumann ranges is the *ground state range*. It is natural to expect that the ground state range consists of one point.

Theorem 2.45. *With $f(u)$ satisfying the conditions (2.211) and (2.226), assume that the solution set of the Neumann problem (2.212) consists of a single curve. Then the set of $\alpha = u(0)$, giving ground state solutions (i.e., solutions of (2.223)), is either one point or an interval.*

Proof: Since the Dirichlet curve extends to infinity, and continues globally for both decreasing and increasing values of $u(0)$, it follows that the set of $\alpha = u(0)$ for which solutions of (2.219) become eventually zero, is an interval (d, ∞), with some $d > 0$. Similarly, the set of $\alpha = u(0)$, giving the Neumann range is (b, n), with some $n \le d$. If $n = d$, we conclude that $u(0) = n$ gives rise to a unique ground state, otherwise $[d, n]$ produces an interval of ground states. $\diamond$

If there is an interval of α's leading to ground states $u(r, \alpha)$ (i.e., solutions of (2.223)), then $w \equiv u_\alpha(r, \alpha)$ satisfies

$$w''(r) + \tfrac{n-1}{r} w'(r) + f'(u)w(r) = 0, \quad 0 < r < \infty, \qquad (2.228)$$
$$w'(0) = 0, \quad w(0) = 1, \quad w(\infty) = 0.$$

Similarly, if there is a turn on a curve of solutions of Neumann problem (2.212), then the corresponding linearized problem (2.213) has a non-trivial solution, i.e., after rescaling

$$w''(r) + \tfrac{n-1}{r} w'(r) + f'(u)w(r) = 0, \quad 0 < r < \lambda^2, \qquad (2.229)$$
$$w'(0) = 0, \quad w(0) = 1, \quad w'(\lambda^2) = 0,$$

for some $\lambda > 0$. Both of these possibilities (i.e., intervals of ground states, and Neumann branches with turns) can usually be ruled out by the same set of numerical computations, thus providing a computer assisted proof of uniqueness of ground state solutions, as we explain next.

Ruling out the possibility of turns on Neumann curves implies that the Neumann curve bifurcating from $u = b$ is unique by the Theorem 2.44, and hence by Theorem 2.45, the set of α's, giving rise to ground states, is either a point or an interval. And it is one point, once we rule out the case of an interval. To rule out an interval of ground states, we need to show that the problem (2.228) has no solution. Our crucial observation is the following: not only we do not need to know $w(r)$ all the way to infinity, we do not need to compute $w(r)$ too far. Namely, as soon as $u(r)$ enters the region where $f'(u(r)) < 0$, if at some point r_0 we have $w(r_0) < 0$ and $w'(r_0) < 0$, then we have $w(r) < 0$ and $w'(r) < 0$ for all $r > r_0$, since $w(r)$ cannot have points of local negative minimums, as is clear from the equation (2.228) (if $\bar{r}$ is such a point, then $w'(\bar{r}) = 0$, $w''(\bar{r}) \geq 0$, and $f'(u(\bar{r}))w(\bar{r}) > 0$, a contradiction.)

Example $f(u) = -u + u^3 + 2u^4$, $n = 3$. We wish to prove uniqueness of the ground state, i.e., of solution to

$$u'' + \tfrac{2}{r}u' - u + u^3 + 2u^4 = 0, \quad r \in (0, \infty) \qquad (2.230)$$

$$u(r) > 0, \ u'(r) < 0$$

$$u'(0) = u(\infty) = 0.$$

We have computed the Dirichlet (top) and the Neumann (bottom) curves for this problem, see the Figure 2.11. The picture leaves no doubt about the uniqueness of ground state solution, separating the Dirichlet and Neumann ranges, but we continue with the computer assisted proof. We have a Neumann curve bifurcating from $u \equiv b \simeq 0.657$, and which extends above $u(0) = 3.3$. Similarly, the Dirichlet curve extends below $u(0) = 3.4$. This leaves us with the interval $(3.3, 3.4)$, where intervals of ground states, or Neumann branches with a turn, might conceivably happen. (Computations show that the Neumann range goes above 3.3 and the Dirichlet range goes below 3.4, but we leave some "safety", so that the numbers 3.3 and 3.4 could be validated, by using estimates and continuity arguments.)

We took $\alpha = 3.35$ in the middle of the range, and computed the solution $u(r)$ of (2.230), and then solved the corresponding variational equation

$$w''(r) + \tfrac{2}{r}w'(r) + (-1 + 3u^2(r) + 8u^3(r))w(r) = 0, \qquad (2.231)$$

$$w(0) = 1, \ w'(0) = 0$$

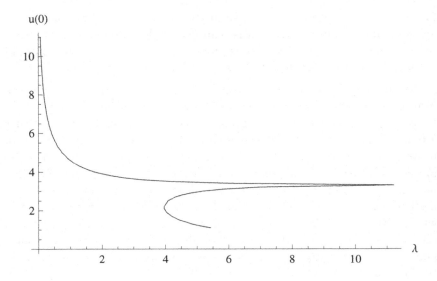

Fig. 2.11   The Neumann and the Dirichlet curves for the equation in (2.230)

on the interval $r \in [0, 1.6]$. The result is plotted in the Figure 2.12. Using the arguments outlined above, P. Korman and Y. Li [107] showed that this picture is sufficient to rule out the possibility of both Neumann branches with a turn, and of intervals of ground states (here $w(1.6) \simeq -0.229$, $w'(1.6) \simeq -0.0587$, and $f'(u(1.6)) < 0$, hence $w'(r) < 0$ for all $r > 1.6$).

Looking at the Figure 2.12, one understands why uniqueness of ground state is tough (and, in fact, it is not known) for the class of functions

$$f(u) = -u^p + au^q + bu^r, \text{ with constants } 1 < p < q < r < \tfrac{n+2}{n-2},$$

which includes the problem (2.230). One needs to prove that $w(\infty) \neq 0$, but for a while $w(r)$ is negative, small in absolute value, and increasing. (Uniqueness of ground states is known in case the function $\frac{uf'(u)}{f(u)}$ is decreasing, see M.K. Kwong and L. Zhang [122], or T. Ouyang and J. Shi [145].)

2.16   Global solution curves for a class of elliptic systems

Following a recent paper of P. Korman [100], we show that bifurcation theory approach applies to systems of Hamiltonian equations on balls, with

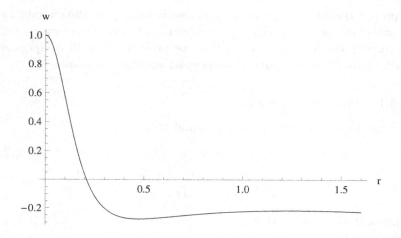

Fig. 2.12   Solution of the variational problem (2.231)

Dirichlet boundary conditions

$$\Delta u + \lambda H_v(u, v) = 0, \quad \text{for } |x| < 1, \quad u = 0 \text{ for } |x| = 1 \qquad (2.232)$$
$$\Delta v + \lambda H_u(u, v) = 0, \quad \text{for } |x| < 1, \quad v = 0 \text{ for } |x| = 1,$$

where $H(u, v) \in C^2(\bar{R}_+ \times \bar{R}_+)$ is a given function, and λ is a positive parameter (we may take the unit ball, by scaling). We shall assume that the partials H_{vv} and H_{uu} are positive, i.e., this system is of "cooperative" type. Then by W.C. Troy's [184] generalization of the classical theorem of B. Gidas, W.-M. Ni and L. Nirenberg [63], any positive solution is radially symmetric, i.e., $u = u(r)$ and $v = v(r)$, $r = |x|$, and moreover $u'(r) < 0$, $v'(r) < 0$, so that $u(0)$ and $v(0)$ give the maximal values of respective functions. We shall also assume that $H_{uv} \leq 0$. Then by a result of P. Korman and J. Shi [119] (which was a generalzation of earlier similar results of R. Dalmasso [43], and P. Korman [91]), the value of $u(0)$ alone uniquely identifies the solution triple $(\lambda, u(r), v(r))$. We shall show that positive solutions of (2.232) lie on global solution curves, i.e., at any point either the Implicit Function Theorem, or the Crandall-Rabinowitz bifurcation Theorem 1.7 applies. By the above remarks, these solution curves can be faithfully represented by two-dimensional curves in $(\lambda, u(0))$ plane. Then, we identify a special class of systems, for which one can write down solution of the linearized system at any singular solution (this includes the turning points). As an application, we get a new non-degeneracy result for power nonlinearities, and a simple proof of uniqueness. In the next chapter, we present considerably more detailed results for the one-dimensional case. We show (in case $n = 1$) that any non-trivial solution of the corresponding linearized problem can be assumed to be positive (in both components), which implies several uniqueness and exact multiplicity results.

2.16.1   *Preliminary results*

We consider a linear system, for $w(r)$ and $z(r)$, on a unit ball

$$w''(r) + \frac{n-1}{r} w'(r) + a(r)w + b(r)z = 0, \quad 0 < r < 1 \qquad (2.233)$$
$$z'' + \frac{n-1}{r} z' + c(r)w + d(r)z = 0, \quad 0 < r < 1$$
$$w'(0) = z'(0) = w(1) = z(1) = 0.$$

Lemma 2.37. *Assume that the given continuous functions $a(r)$, $b(r)$, $c(r)$ and $d(r)$ satisfy*

$$a(r) \leq 0, \; d(r) \leq 0, \; b(r) > 0, \; c(r) > 0 \; \text{for all } r \in [0, 1). \qquad (2.234)$$

Then for any non-trivial solution of (2.233), $z'(1)$ and $w'(1)$ are both non-zero, and have the same sign.

Proof: We divide the proof into three steps.

Step 1. We claim that $z'(1)$ and $w'(1)$ cannot be of the opposite sign. Assume, on the contrary, that $w'(1) < 0$, while $z'(1) > 0$. By continuity,

$$w'(r) < 0, \; \text{and} \; z'(r) > 0 \; \text{for } r \text{ close to 1.} \qquad (2.235)$$

Let $r_0 \geq 0$ denote the infimum of r's, for which (2.235) holds. We shall assume that $w'(r_0) = 0$, since the other case, when $z'(r_0) = 0$, is similar. Observe that $w''(r_0) \leq 0$ (the opposite inequality $w''(r_0) > 0$ together with $w'(r_0) = 0$, would imply that $w'(r) > 0$ to the right of r_0, contradicting (2.235)). From (2.235), and the boundary conditions, we also conclude that $w(r_0) > 0$ and $z(r_0) < 0$. Hence, the left hand side of the first equation in (2.233) is negative at r_0, a contradiction, proving that (2.235) holds for all $r \geq 0$. But that contradicts the initial conditions at $r = 0$.

Step 2. Claim: if (w, z) is a non-trivial solution of (2.233), then we may assume that $w(r)$ and $z(r)$ are both positive near $r = 1$.

To prove that, we begin by noticing that by uniqueness for initial value problems, $w'(1)$ and $z'(1)$ cannot be both zero. We may assume that $w'(1) < 0$, and then by the Step 1, $z'(1) \leq 0$. By continuity, $w(r)$ and $-w'(r)$ are positive near $r = 1$, say on the interval $(\beta, 1)$, for some β. If

$z(r)$ failed to be positive near $r = 1$, then there are two possibilities. The first possibility is that $z'(1) = 0$ and $z(r) < 0$ on some interval $(\beta_1, 1)$, with $\beta \leq \beta_1$. Then we can find a point $r_1 \in (\beta_1, 1)$, such that $w(r_1) > 0$, $w'(r_1) < 0$, and $z(r_1) < 0$, $z'(r_1) > 0$. As above, we show that these inequalities imply that $w'(r) < 0$ and $z'(r) > 0$ for all $r \in [0, r_1]$, with a contradiction at $r = 0$. The second possibility is that $z(r)$ has infinitely many roots on the interval $(\beta, 1)$. But then we have a contradiction in the second equation of (2.233), at any point of negative minimum of $z(r)$ (all of the terms on the left in (2.233) are either positive or non-negative).

Step 3. Since $z(r)$ is positive near $r = 1$, we have $w''(r) + \frac{n-1}{r} w'(r) + a(r)w \leq 0$ near $r = 1$, and hence $w'(1) < 0$, by Hopf's boundary lemma (see e.g., p. 107 in [157]). Similarly, $z'(1) < 0$. $\Diamond$

Lemma 2.38. *Under the conditions (2.234), the solution space of the system (2.233) is one-dimensional (i.e., solution (w, z) is a constant multiple of any other solution $(\bar{w}, \bar{z})$).*

Proof: If we denote $p = w'(1)$ and $q = z'(1)$, then the vector (p, q) uniquely identifies any solution of the system (2.233) (i.e., the system (2.233) has at most one solution, with these initial conditions). For any constant t, the vector (tp, tq) uniquely identifies the constant multiple of the same solution. If the vector (p_1, q_1) gives rise to another solution of (2.233), then the same is true for the vector $(p + p_1, q + q_1)$, by linearity. So, if the set of vectors (p, q), giving rise to solutions of (2.233), is not one-dimensional, it is all of R^2. But the vector $(1, -1)$ does not belong to that set by Lemma 2.37, a contradiction. $\Diamond$

Remark In general, the solution space of (2.233) may be two-dimensional, even in case $n = 1$, as we shall see in the next chapter.

2.16.2   *Global solution curves for Hamiltonian systems*

Given a function $H(u, v) \in C^2(\bar{R}_+ \times \bar{R}_+)$, we consider positive solutions (i.e., $u > 0$ and $v > 0$) of the system (2.232). We shall assume that

$$H_{uu} > 0, \quad \text{and} \quad H_{vv} > 0 \quad \text{for all } (u, v) > 0. \qquad (2.236)$$

By W.C. Troy's [184] extension of the classical results of B. Gidas, W.-M. Ni and L. Nirenberg [63], under the condition (2.236), any positive solution of (2.232) is radially symmetric, i.e., $u = u(r)$ and $v = v(r)$, $r = |x|$, and so (2.232) becomes a system of ODE's

$$u'' + \frac{n-1}{r}u' + \lambda H_v(u,v) = 0, \quad \text{for } r < 1, \quad u'(0) = u(1) = 0 \quad (2.237)$$
$$v'' + \frac{n-1}{r}v' + \lambda H_u(u,v) = 0, \quad \text{for } r < 1, \quad v'(0) = v(1) = 0.$$

We shall also assume that

$$H_{uv} \le 0, \quad \text{for all } (u,v) \ge 0. \tag{2.238}$$

According to P. Korman and J. Shi [119] (see also R. Dalmasso [43], and P. Korman [91]), under the conditions (2.236) and (2.238), the value of $u(0)$ alone (or the value of $v(0)$) uniquely identifies the solution triple $(\lambda, u(r), v(r))$, i.e., the value of $u(0)$ can be used as a global parameter on the solution curves of (2.237). We shall need the linearized problem, corresponding to (2.237)

$$w'' + \frac{n-1}{r}w' + \lambda H_{uv}w + \lambda H_{vv}z = 0, \quad \text{for } r < 1, \tag{2.239}$$
$$z'' + \frac{n-1}{r}z' + \lambda H_{uu}w + \lambda H_{uv}z = 0, \quad \text{for } r < 1,$$
$$w'(0) = w(1) = 0, \quad z'(0) = z(1) = 0.$$

We call a solution (u,v) of (2.237) *singular*, if the linearized system (2.239) has a non-trivial solution. By Lemma 2.38, the solution set of (2.239) is one-dimensional, under our assumptions.

Lemma 2.39. *Assume that a positive solution (u,v) of (2.237) is singular, i.e., the linearized system (2.239) has a non-trivial solution (w,z). Then*

$$\int_0^1 (H_v(u,v)z + H_u(u,v)w)\, r^{n-1}\, dr = \frac{1}{2\lambda}\left(v'(1)w'(1) + u'(1)z'(1)\right).$$

Proof: One easily checks that the functions $p(r) = r u_r$, $q(r) = r v_r$ satisfy the system

$$p'' + \frac{n-1}{r}p' + \lambda H_{vu}p + \lambda H_{vv}q = -2\lambda H_v, \tag{2.240}$$
$$q'' + \frac{n-1}{r}q' + \lambda H_{uu}p + \lambda H_{uv}q = -2\lambda H_u.$$

We multiply the first equation in (2.239) by q, and subtract from that the first equation in (2.240), multiplied by z. The result can be put into the form

$$(qw' - zp')' - q'w' + z'p' + \frac{n-1}{r}(qw' - zp') \tag{2.241}$$
$$+ \lambda H_{vu}wq - \lambda H_{vu}zp = 2\lambda H_v z.$$

Similarly, we multiply the second equation in (2.239) by p, and subtract from that the second equation in (2.240), multiplied by w. The result can be put into the form

$$(pz' - wq')' - z'p' + q'w' + \frac{n-1}{r}(pz' - wq') \tag{2.242}$$
$$+ \lambda H_{uv}zp - \lambda H_{uv}qw = 2\lambda H_u w.$$

We now add the equations (2.241) and (2.242), and put the result in the form

$$\left[r^{n-1}(qw' - zp')\right]' + \left[r^{n-1}(pz' - wq')\right]' = r^{n-1}(2\lambda H_v z + 2\lambda H_u w).$$

Integrating this over the interval $(0, 1)$, we conclude the proof. $\diamond$

Positive solutions of the system (2.237) satisfy $u'(1) \leq 0$, $v'(1) \leq 0$. We shall assume that one of these derivatives is non-zero:

$$|u'(1)| + |v'(1)| > 0. \tag{2.243}$$

For concrete systems this condition can be usually justified, by using Hopf's boundary lemma.

Theorem 2.46. *Assume that the conditions (2.236), (2.238), and (2.243) are satisfied for the system (2.232). Then all positive solutions of (2.232) lie on global continuous non-intersecting curves in $(\lambda, u(0))$ plane (or in $(\lambda, v(0))$ plane).*

Proof: By the above remarks, any positive solution of the system (2.232) is radially symmetric, and hence we may consider the system (2.237). We show that at any positive solution of (2.237) either the Implicit Function Theorem, or the Crandall-Rabinowitz bifurcation Theorem 1.7 applies, and hence we can always continue the solutions (that is what we mean by "the global curves"). To recast our system in the operator form, we define the spaces, $\mathbf{X} = \{(u(r), v(r)) \mid u, v \in C^{2,\alpha}[0, 1), u'(0) = u(1) = v'(0) = v(1) = 0\}$ and $\mathbf{Y} = \{(u(r), v(r)) \mid u, v \in C^{\alpha}[0, 1)\}$, and consider the map

$$F(u, v, \lambda) = \begin{pmatrix} u'' + \frac{n-1}{r}u' + \lambda H_v(u, v) \\ v'' + \frac{n-1}{r}v' + \lambda H_u(u, v) \end{pmatrix} : \mathbf{X} \to \mathbf{Y}.$$

Clearly, our system (2.237) can be recast in the operator form $F(u, v, \lambda) = 0$, and the system (2.239) gives the corresponding linearized problem.

We claim that the linearized operator $F_{(u,v)}(p, q)$ is a Fredholm operator of index zero. This can be seen by recasting the linearized equations as a system of two integral equations, and then the solution operator defines a compact operator on the product space $\mathbf{Y} \to \mathbf{Y}$, so that we can use the Riesz-Schauder theory of compact operators. If (2.239) has only the trivial solution, then $F_{(u,v)}(p,q)$ is one-to-one, and hence onto, and the Implicit Function Theorem applies, and we can continue the solution to nearby λ's. Assume now that at some $(\bar{\lambda}, \bar{u}, \bar{v})$ the system (2.239) has a non-trivial solution. Then, by Lemma 2.38, the solution set of (2.239), i.e., the null-space of $F_{(u,v)}(\bar{\lambda}, \bar{u}, \bar{v})(p,q)$ is one-dimensional. Since $F_{(u,v)}(\bar{\lambda}, \bar{u}, \bar{v})(p,q)$ is a Fredholm operator of index zero, the range of the map $F_{(u,v)}(\bar{\lambda}, \bar{u}, \bar{v})(p,q)$ has codimension one. To apply the Crandall-Rabinowitz Theorem 1.7, it remains to show that $F_\lambda(\bar{\lambda}, \bar{x}) \notin R(F_x(\bar{\lambda}, \bar{x}))$, with $\bar{x} = (\bar{u}, \bar{v})$. Assuming otherwise, there would exist $(W(x), Z(x)) \in \mathbf{X} \times \mathbf{X}$, so that

$$W'' + \tfrac{n-1}{r}W' + \lambda H_{uv}W + \lambda H_{vv}Z = H_v \qquad (2.244)$$
$$Z'' + \tfrac{n-1}{r}Z' + \lambda H_{uu}W + \lambda H_{uv}Z = H_u$$
$$W'(0) = W(1) = 0, \quad Z'(0) = Z(1) = 0\,.$$

Proceeding the same way as we did in the proof of Lemma 2.39, we get

$$\int_0^1 \left(H_v(u,v)z + H_u(u,v)w \right) r^{n-1}\, dr = 0\,.$$

However, by Lemmas 2.37, 2.39, and our assumptions, this integral is not zero, a contradiction. $\qquad\qquad\qquad\qquad\qquad\qquad\qquad\qquad\quad \diamond$

2.16.3 *A class of special systems*

Following P. Korman [100], we now identify a special class of systems, for which one can write down a solution of the linearized system at any singular solution. This is similar to the case of functions $f(u) = e^u$ and $f(u) = (u + \gamma)^p$ in the scalar case, that we considered before. Unlike the scalar case, these solutions do not necessarily satisfy the boundary conditions (since we can scale only one of the unknown functions, to take a desired value). Still, for power nonlinearities we can prove that any positive solution is non-singular, giving an easy approach to uniqueness.

We consider positive solutions of the system

$$u'' + \tfrac{n-1}{r}u' + f(v) = 0, \quad \text{for } r < 1, \ u'(0) = u(1) = 0 \qquad (2.245)$$
$$v'' + \tfrac{n-1}{r}v' + g(u) = 0, \quad \text{for } r < 1, \ v'(0) = v(1) = 0\,,$$

which is a subclass of the Hamiltonian system (2.237). (Here $H(u,v) = F(v) + G(u)$, where F and G are anti-derivatives of f and g respectively.)

To be consistent with the conditions of the Theorem 2.46, we shall assume that

$$f(0) \geq 0, \quad g(0) \geq 0, \quad \text{and} \quad f'(t) > 0, g'(t) > 0, \text{ for all } t > 0. \quad (2.246)$$

The linearized problem, corresponding to (2.245), is

$$w'' + \tfrac{n-1}{r}w' + f'(v)z = 0, \quad \text{for } r < 1, \quad w'(0) = w(1) = 0 \quad (2.247)$$
$$z'' + \tfrac{n-1}{r}z' + g'(u)w = 0, \quad \text{for } r < 1, \quad z'(0) = z(1) = 0.$$

We shall also consider a linear problem

$$p'' + \tfrac{n-1}{r}p' + f'(v)q = 0, \quad \text{for } r < 1, \quad p'(0) = 0 \quad (2.248)$$
$$q'' + \tfrac{n-1}{r}q' + g'(u)p = 0, \quad \text{for } r < 1, \quad q'(0) = 0,$$

which is different from (2.247), as we do not impose the boundary conditions at $r = 1$.

We now identify a class of systems, for which it is possible to write down explicitly the solution of the problem (2.248).

Theorem 2.47. *Assume that (u, v) is any positive solution of (2.245). Assume that the functions $f(t)$ and $g(t)$ are both of one of the two forms ce^{at} or $c(t + b)^p$, with some positive constants a,b,c and p (i.e., any one of the four possible combinations holds). In case $f(t) = c(t + b)^{p_1}$ and $g(t) = c(t + b_1)^{p_2}$, we assume additionally that $p_1 p_2 \neq 1$. Then there is a simple formula, expressing a solution (w, z) of (2.248) through (u, v).*

Proof: Let $p(r) = ru'(r) + \mu_1 u(r) + \alpha$, and $q(r) = rv'(r) + \mu_2 v(r) + \beta$. These function satisfy

$$p'' + \tfrac{n-1}{r}p' + f'(v)q = -2f - \mu_1 f + \mu_2 vf' + \beta f' \quad (2.249)$$
$$q'' + \tfrac{n-1}{r}q' + g'(u)p = -2g - \mu_2 g + \mu_1 ug' + \alpha g'.$$

Our goal is to choose the constants μ_1, μ_2, α and β, to make both quantities on the right to be zero. If both f and g are exponentials, $f(v) = ae^{\gamma v}$, and $g(u) = be^{\delta u}$, we take $\mu_1 = \mu_2 = 0$, $\alpha = \frac{2}{\delta}$ and $\beta = \frac{2}{\gamma}$. In the "mixed" case, when $f(v) = ae^{\delta v}$, and $g(u) = b(u + \gamma)^p$, we select $\mu_2 = 0$, then $\mu_1 = \frac{2}{p}$ and $\alpha = \frac{2\gamma}{p}$, followed by $\beta = \frac{2+\mu_1}{\delta}$. Finally, if both f and g are powers, $f(v) = a(v + \gamma)^{p_1}$ and $g(u) = b(u + \delta)^{p_2}$, we set $\beta = \mu_2 \gamma$, and $\alpha = \mu_1 \delta$. We then need to select μ_1 and μ_2, so that

$$-\mu_1 + p_1 \mu_2 = 2$$
$$p_2 \mu_1 - \mu_2 = 2,$$

i.e., $\mu_1 = \frac{2p_1+2}{p_1p_2-1}$, and $\mu_2 = \frac{2p_2+2}{p_1p_2-1}$. ◇

Remark It is natural to ask if $(w, z) = (p, q)$. In case of one equation such result follows by scaling, but for systems, only one of the unknown functions can be scaled to take a desired value.

We now turn to the power nonlinearities.

Theorem 2.48. *Assume that $f(v) = v^{p_1}$ and $g(u) = u^{p_2}$, with $p_1p_2 \neq 1$. Then any positive solution of (2.245) is non-singular, i.e., the corresponding linearized problem (2.247) has only the trivial solution $w = z = 0$.*

Proof: Assume, on the contrary, that (w, z) is a non-trivial solution of (2.247). By the Theorem 2.47, we can find a solution of the problem (2.248), in the form $p = ru_r + \mu_1 u$, and $q = rv_r + \mu_2 v$. It follows by Hopf's boundary lemma that

$$p(1) = u'(1) < 0, \quad q(1) = v'(1) < 0. \tag{2.250}$$

By scaling (w, z), we can achieve $w(0) = p(0)$. We then have $z(0) \neq q(0)$ (otherwise by uniqueness for initial value problems, $(w, z) = (p, q)$, but the pair (p, q) does not satisfy the boundary condition at $r = 1$). Assume, for definiteness, that $z(0) > q(0)$. Call $P = p - w$, and $Q = q - z$. Then the pair (P, Q) satisfies the same linear system (2.247), which we write in the form

$$(r^{n-1}P')' = -r^{n-1}f'(v)Q, \quad P(0) = 0, \; P'(0) = 0 \tag{2.251}$$
$$(r^{n-1}Q')' = -r^{n-1}g'(u)P, \quad Q(0) < 0, \; Q'(0) = 0.$$

By continuity, for r small

$$Q(r) < 0. \tag{2.252}$$

It follows from the first equation in (2.251) that the function $r^{n-1}P'$ is positive, and so $P(r)$ is positive and increasing, i.e., for small r

$$P(r) > 0. \tag{2.253}$$

But then, integrating the second equation in (2.251), we see that $Q'(r) < 0$, i.e., $Q(r)$ is decreasing, and so the inequality (2.252) continues to hold. We see that the inequalities (2.252) and (2.253) "reinforce" each other, so that they both hold for $r \in (0, 1]$. At $r = 1$, we have

$$0 < P(1) = p(1) - w(1) \leq -w(1),$$

i.e., $w(1) < 0$, a contradiction. ◇

As an application, we can give a complete description of the solution set for the case of power nonlinearities.

Theorem 2.49. *Consider the system*

$$\Delta u + \lambda v^{p_1} = 0, \quad for \; |x| < 1, \quad u = 0 \; for \; |x| = 1 \quad (2.254)$$
$$\Delta v + \lambda u^{p_2} = 0, \quad for \; |x| < 1, \quad v = 0 \; for \; |x| = 1,$$

with given positive constants p_1 and p_2, satisfying $p_1 p_2 \neq 1$, and λ a positive parameter. Then all positive solutions of (2.254) are radially symmetric, and they lie on a unique continuous curve in $(\lambda, u(0))$ plane (or in $(\lambda, v(0))$ plane). Moreover, all solutions are non-singular, and so the solution curve admits no turns. In the sublinear case, when $p_1 p_2 < 1$, there is a curve of solutions, starting at $\lambda = 0$, $u(0) = 0$, and continuing without any turns to $\lambda = \infty$, $u(0) = \infty$, with both components of $(\lambda, u(0))$ strictly increasing. In the superlinear case, when $p_1 p_2 > 1$, if the problem (2.254) has a positive solution at some λ, then all solutions lie on a unique curve of solutions, on which λ is increasing and $u(0)$ is decreasing, and $u(0) \to 0$, as $\lambda \to \infty$, while $u(0) \to \infty$, as $\lambda \to 0$.

Proof: In the sublinear case, there exists a positive solution for any λ, see e.g., R. Dalmasso [43], or P. Korman [87]. In the superlinear case, some restrictions on p_1 and p_2 are necessary, see e.g., L.A. Peletier and R.C.A.M. Van der Vorst [150], and P. Clément, D.G. de Figueiredo and E. Mitidieri [36], for the conditions on p_1 and p_2 ensuring existence of solutions (we review them below). By the Theorem 2.48, any solution of (2.254) is non-singular, and hence by the Implicit Function Theorem, we can continue solutions in λ, on a solution curve, which does not turn. By the result of P. Korman and J. Shi [119], mentioned above, the value of $u(0)$ changes monotonously on this curve. The rest follows by a simple scaling. Indeed, if we let $u = \alpha U$ and $v = \beta V$, with $\alpha^{\frac{p_1 p_2 - 1}{p_1 + 1}} = \frac{1}{\lambda}$ and $\beta = \alpha^{\frac{p_2 + 1}{p_1 + 1}}$, we see that (U, V) satisfies the problem (2.254), with $\lambda = 1$. It follows that in the sublinear (superlinear) case solutions tend to infinity (zero) as $\lambda \to \infty$, and to zero (infinity) as $\lambda \to 0$. Since any solution curve "takes up" all possible values of $u(0)$, it follows that there is only one solution curve. $\diamond$

Remarks

(1) The result is not true in case $p_1 p_2 = 1$. In fact, if $p_1 p_2 = 1$, and the problem (2.254) has a non-trivial solution (u, v), then (2.254) has a continuum of solutions of the form $(\alpha u, \alpha^{1/p_1} v)$, for any $\alpha > 0$. Pre-

sumably, in this case the problem has non-trivial solutions only for a sequence of eigenvalues. In case $p_1 = p_2 = 1$, this is not hard to prove.

(2) Since solutions are non-singular, they persist under small perturbations of the system.

In the superlinear case, when $1 \leq p_1, p_2 \leq \frac{n+2}{n-2}$, $p_1 p_2 \neq 1$, a notion of *critical hyperbola* was introduced in P. Clément, D.G. de Figueiredo and E. Mitidieri [36], and L.A. Peletier and R.C.A.M. Van der Vorst [150]:

$$\frac{1}{p_1 + 1} + \frac{1}{p_2 + 1} = 1 - \frac{2}{n}.$$

It extends the notion of the critical exponent from scalar case. It turns out that the problem (2.254) is solvable in the *subcritical* case, i.e., when

$$\frac{1}{p_1 + 1} + \frac{1}{p_2 + 1} > 1 - \frac{2}{n}.$$

The above theorem describes the solution curve on a ball in that case.

2.17 The case of a "thin" annulus

On an annulus $\Omega = \{x \mid A < |x| < B\}$ in R^n, $n \geq 2$, we consider the problem

$$\Delta u + \lambda f(u) = 0 \text{ in } \Omega, \quad u = 0 \text{ on } \partial\Omega. \tag{2.255}$$

We study positive radially symmetric solutions of (2.255), depending on a positive parameter λ. It is known that the problem (2.255) may have positive non-radial solutions, in contrast to the case when domain is a ball in R^n, and all positive solutions are necessarily radially symmetric. This problem arises in many applications, and there is a large literature on the subject, including W.-M. Ni and R. Nussbaum [138], S.S. Lin [128], H. Aduén, A. Castro, and J. Cossio [5]. To get exact multiplicity results, we shall restrict our attention to the case of "thin" annulus, which we define next. Set $c_n = (2n - 3)^{\frac{1}{n-2}}$ for $n \geq 3$, and $c_2 = e^2$. We shall assume that

$$B \leq c_n A. \tag{2.256}$$

The special role of "thin" annulus was recognized first by W.-M. Ni and R. Nussbaum [138], but they had a more restrictive condition, involving $c_n = (n - 1)^{\frac{1}{n-2}}$ for $n \geq 3$, and $c_2 = e$. Their condition appeared later in S.S. Lin [128]. The condition (2.256), and the results of this section are taken from P. Korman [83].

The crucial part of our study is to prove positivity of the radial solutions of the linearized problem

$$\Delta w + \lambda f'(u)w = 0 \text{ in } \Omega, \quad w = 0 \text{ on } \partial\Omega. \tag{2.257}$$

In fact, if $f(u) \in C^2(\bar{R}_+)$ satisfies

$$f(u) > 0 \quad \text{for all } u > 0, \tag{2.258}$$

then we showed in [83] that any non-trivial solution of (2.257) is of one sign.

We then obtained uniqueness and exact multiplicity results, following the familiar scheme. The following theorem extends the Theorem 2.4 in W.-M. Ni and R.D. Nussbaum [138], allowing for a larger annulus.

Theorem 2.50. *([83]) Assume that $B \leq c_n A$, $f(u) \in C^2(\bar{R}_+)$, $f(u) > 0$ and $uf'(u) > f(u)$ for all $u > 0$, and assume finally that $\lim_{u \to \infty} \frac{f(u)}{u} = \infty$. Then the problem (2.255) has at most one positive radially symmetric solution for any $\lambda > 0$.*

We have the following exact multiplicity result for convex $f(u)$.

Theorem 2.51. *([83]) Assume that $B \leq c_n A$, $f(u) \in C^2(\bar{R}_+)$, $f(u) > 0$ for $u \geq 0$, and $\lim_{u \to \infty} \frac{f(u)}{u} = \infty$, and assume finally that*

$$f''(u) > 0 \text{ for almost all } u > 0.$$

Then there is a critical $\lambda_0 > 0$, such that the problem (2.255) has exactly two positive radially symmetric solutions for $0 < \lambda < \lambda_0$, it has exactly one positive radially symmetric solution at $\lambda = \lambda_0$, and no positive radially symmetric solutions for $\lambda > \lambda_0$. Moreover, all positive radially symmetric solutions lie on a single smooth solution curve, which for $\lambda \in (0, \lambda_0)$ has two branches $u^-(r, \lambda)$ and $u^+(r, \lambda)$, with $u^-(r, \lambda) < u^+(r, \lambda)$ for all $r = |x| \in (A, B)$. The lower branch $u^-(r, \lambda)$ is strictly monotone increasing in λ, and $\lim_{\lambda \to 0+} u^-(r, \lambda) = 0$, for all $r \in (A, B)$. For the upper branch, $\lim_{\lambda \to 0+} \max_r u^+(r, \lambda) = \infty$.

Proofs of these results follow the familiar pattern, once positivity of the radial solutions of the linearized problem is established, which we discuss next.

Radial solutions of (2.255) satisfy

$$u'' + \frac{n-1}{r}u' + \lambda f(u) = 0 \quad \text{for } A < r < B, \quad u(A) = u(B) = 0. \tag{2.259}$$

We make a standard change of variables. In case $n \geq 3$, we let $s = r^{2-n}$ and $u(s) = U(r)$, transforming (2.259) into the problem

$$u'' + \alpha(s)f(u) = 0, \quad \text{for } a < s < b, \ u(a) = u(b) = 0, \tag{2.260}$$

where $\alpha(s) = (n-2)^{-2}s^{-2k}$, with $k = 1 + \frac{1}{n-2}$, $a = B^{2-n}$ and $b = A^{2-n}$. In case $n = 2$, we set $s = -\log r$, and $u(s) = U(r)$, obtaining again the problem (2.260), this time with $\alpha(s) = e^{-2s}$, and $a = -\log B$, $b = -\log A$. Corresponding linearized problem is

$$w'' + \alpha(s)f'(u)w = 0, \quad \text{for } a < s < b, \ w(a) = w(b) = 0. \tag{2.261}$$

Theorem 2.52. *Assume that the annulus is thin, i.e., the condition (2.256) holds, and the function $f(u) \in C^1(\bar{R}_+)$ satisfies $f(u) > 0$ for $u > 0$. Then any non-trivial solution of (2.261) is of one sign.*

We postpone the proof to the next chapter, where it will be a corollary of more general results.

2.18   A class of p-Laplace problems

We consider positive solutions of the problem

$$\Delta_p u + f(|x|, u) = 0, \quad \text{for } |x| < 1, \ u = 0 \text{ on } |x| = 1, \tag{2.262}$$

where Δ_p denotes the p-Laplace operator

$$\Delta_p u = \text{div}\left(|\nabla u|^{p-2}\nabla u\right), \quad p > 1.$$

We have discussed a number of uniqueness and exact multiplicity results in case $p = 2$, i.e., for the classical Laplacian. When $p \neq 2$, there are a number of additional technical difficulties, some of which were overcome only recently, through the efforts of many of people. In particular, F. Brock [24] and L. Damascelli and F. Pacella [45] have extended the symmetry results of B. Gidas, W.-M. Ni and L. Nirenberg [63]. J. Serrin and H. Zou [168] have proved a Liouville type result, which implies a priori estimates of B. Gidas and J. Spruck type [64], provided that the point of maximum of solution is bounded away from the boundary, as was observed in A. Aftalion and F. Pacella [11], and J. Fleckinger and W. Reichel [57]. A. Aftalion and F. Pacella [11] pointed out the appropriate function spaces, in order to apply the Implicit Function Theorem. In fact, A. Aftalion and F. Pacella [11] have proved uniqueness of solution to (2.262), for $f(r, u)$ similar to the one we consider below. In this section, following P. Korman

[101], we provide a simpler proof of the crucial step in [11], involving the non-degeneracy of solutions. Moreover, we clarify the proper conditions, and provide the existence part for the model examples.

We assume that $f(r, u) \in C^1([0, 1] \times \bar{R}_+)$ is a continuously differentiable function (with $r = |x|$), and

$$f(r, 0) = 0, \quad \text{and} \quad f_r(r, u) \leq 0, \quad \text{for all } r \in [0, 1], \text{ and } u > 0. \quad (2.263)$$

If $1 < p \leq 2$, then in view of B. Gidas, W.-M. Ni and L. Nirenberg [63] and L. Damascelli and F. Pacella [45], positive solutions of (2.262) are radially symmetric, and hence they satisfy

$$\varphi(u'(r))' + \tfrac{n-1}{r}\varphi(u'(r)) + f(r, u(r)) = 0, \quad 0 < r < 1, \quad (2.264)$$
$$u'(0) = u(1) = 0.$$

where we denote $\varphi(t) = t|t|^{p-2}$. In case $p > 2$, where no symmetry result is available, we just restrict our attention to the radial solutions of (2.262), i.e., we again consider (2.264). Following B. Franchi et al [59], we consider *classical* solutions of (2.264), i.e., we assume that $u \in C^1[0, 1]$ and $\varphi(u') \in C^1(0, 1]$. As a consequence, if $u'(r_1) = 0$ for some $r_1 \in (0, 1)$, we may define by continuity $\varphi(u'(r_1)) = 0$.

The following lemma gives a known condition for the Hopf's lemma to hold, see J.L. Vázquez [185], or A. Aftalion and F. Pacella [11].

Lemma 2.40. *Assume that* $f(r, u) \in C^1([0, 1] \times \bar{R}_+)$ *satisfies* $f(r, 0) = 0$, *for all* $r \in [0, 1]$, *and either* f *is non-negative, or for some* $c_0 > 0$

$$-f(r, s) < c_0 s^{p-1}, \quad \text{for small } s > 0, \text{ and } r \text{ near } 1. \quad (2.265)$$

If $u(r)$ *is a positive solution of (2.264), then*

$$u'(1) < 0. \quad (2.266)$$

The next lemma extends the Proposition 1.2.6 in B. Franchi et al [59], which considered the case of $f = f(u)$, i.e., independent of r. Again, we define $F(r, u) = \int_0^u f(r, t)\, dt$.

Lemma 2.41. *Assume that*

$$F_r(r, u) \leq 0, \quad \text{for all } r \in [0, 1], \text{ and } u > 0; \quad (2.267)$$

for each $r_0 \in [0, 1)$, *the function* $f(r_0, u)$ *is either positive for* $\quad (2.268)$

all $u > 0$, *or it changes sign once from negative, for small* u, *to positive.*

Then any positive solution of (2.264) satisfies

$$u'(r) < 0 \quad \text{for all } r \in (0, 1), \quad (2.269)$$
$$f(0, u(0)) > 0. \quad (2.270)$$

Proof: We begin by showing that $u(r)$ has no local minimums. Indeed, let $r_0 \in [0, 1]$ be a point of local minimum, i.e., $u'(r_0) = 0$, $u''(r_0) \geq 0$. We claim that

$$f(r_0, u(r_0)) \leq 0. \tag{2.271}$$

If $p \geq 2$, this follows by evaluating the equation (2.264) at r_0. In general, we can find a sequence $\{r_n\}$ tending to r_0 from the right, at which $u'(r_n) > 0$, $u''(r_n) \geq 0$. Evaluating the equation (2.264) at r_n, we have $f(r_n, u(r_n)) < 0$, and taking the limit we conclude (2.271). In case f is positive, (2.271) implies a contradiction. In the other case, (2.271) implies that $F(r_0, u(r_0)) < 0$, and hence the "energy" $E(r) \equiv \frac{p-1}{p}|u'(r)|^p + F(r, u(r))$ satisfies $E(r_0) < 0$. We have

$$E'(r) = -\frac{n-1}{r}\varphi(u')u' + F_r(r, u) \leq 0,$$

and so the energy is non-increasing. But $E(1) = \frac{p-1}{p}|u'(1)|^p \geq 0$, a contradiction. We conclude that $u'(r) \leq 0$ for all r. To get the strict inequality, assume that $u'(r_1) = 0$ at some $r_1 \in (0, 1)$. As discussed above, we have $\varphi(u'(r_1)) = 0$, and we also have $\frac{d}{dr}\varphi(u'(r_1)) = 0$, since r_1 is not an extremum. From the equation (2.264), $f(r_1, u(r_1)) = 0$, implying that $E(r_1) < 0$, which results in the same contradiction as before. Finally, we conclude (2.270), by using the same argument one more time, since the opposite inequality would imply $E(0) < 0$. $\diamond$

We assume that $f(r, u)$ satisfies the following two additional conditions, which are similar to the ones in [11]

$$uf_u(r, u) - (p-1)f(r, u) > 0 \quad \text{for all } r \in [0, 1], \text{ and } u > 0; \tag{2.272}$$

For any positive solution of (2.264)

$$\alpha(r) = \frac{pf(r, u(r)) + rf_r(r, u(r))}{u(r)f_u(r, u(r)) - (p-1)f(r, u(r))} \tag{2.273}$$

is a non-increasing function of r, for $r \in (0, 1)$.

We shall need to consider the linearized problem corresponding to (2.264) (here $w = w(r)$, $0 < r < 1$)

$$(\varphi'(u'(r))w'(r))' + \frac{n-1}{r}\varphi'(u'(r))w'(r) + f_u(r, u(r))w(r) = 0, \tag{2.274}$$
$$w'(0) = w(1) = 0.$$

We will show that under the above conditions any positive solution of (2.264) is non-singular, i.e., the problem (2.274) admits only the trivial solution. The following technical lemma we proved in [93].

Lemma 2.42. *Let $u(r)$ be a positive solution of (2.264), and assume that the function $f(r, u)$ satisfies the conditions (2.263) and (2.272). Then the function $f(r, u(r))$ can change sign at most once on $(0, 1)$.*

Proof: Let $\xi \in (0, 1)$ be such that $f(\xi, u(\xi)) = 0$. We claim that $f(r, u(r)) > 0$ for all $r \in [0, \xi]$. Indeed, by (2.272) we conclude that $f_u(\xi, u(\xi)) > 0$, and in general $f_u(r, u(r)) > 0$, so long as $f(r, u(r)) > 0$, and hence

$$\frac{d}{dr} f(r, u(r)) = f_r(r, u(r)) + f_u(r, u(r))u'(r) < 0,$$

and the claim follows. So that, if the function $f(r, u(r))$ is positive near $r = 1$, it is positive for all $r \in [0, 1)$. If, on the other hand, $f(r, u(r))$ is negative near $r = 1$, it will change sign exactly once on $[0, 1)$ (since it cannot stay negative for all r, by the maximum principle). $\diamond$

This lemma implies that either $f(r, u(r)) > 0$ for all $r \in [0, 1)$, or else there is a point $r_2 \in (0, 1)$, so that $f(r, u(r)) > 0$ on $[0, r_2)$ and $f(r, u(r)) < 0$ on $(r_2, 1)$. We put these cases together, we defining $r_2 = 1$ in the first case. I.e., the last lemma implies that

$$\text{there is a } r_2 \in (0, 1], \text{ so that } f(r, u(r)) > 0 \text{ on } [0, r_2) \quad (2.275)$$
$$\text{and, in case } r_2 < 1, \text{ we have: } f(r, u(r)) < 0 \text{ on } (r_2, 1).$$

We shall consider the following two functions, depending on the solutions of (2.264) and (2.274):

$$\xi(r) = r^{n-1} \left[(p-1)\varphi(u'(r))w(r) - \varphi'(u'(r))u(r)w'(r) \right] ; \quad (2.276)$$

$$T(r) = r^n \left[(p-1)\varphi(u'(r))w'(r) + f(r, u(r))w(r) \right] \quad (2.277)$$
$$+ (n-p)r^{n-1}\varphi(u'(r))w(r).$$

In case $p = 2$, these functions were introduced by M. Tang [182]. The following crucial lemma is proved by a direct computation.

Lemma 2.43. *For any solutions of (2.264) and (2.274), we have*

$$\xi'(r) = r^{n-1}w(r) \left[u(r)f_u(r, u(r)) - (p-1)f(r, u(r)) \right] ; \quad (2.278)$$

$$T'(r) = r^{n-1}w(r) \left[pf(r, u(r)) + rf_r(r, u(r)) \right] . \quad (2.279)$$

We shall need two more functions, depending on the solutions of (2.264), see [144], [182], [11], and [93] for similar functions in case $p = 2$,

$$Q(r) = r^n \left[(p-1)\varphi\left(u'(r)\right)u'(r) + u(r)f(r, u(r)) \right] \qquad (2.280)$$
$$+ (n-p)r^{n-1}\varphi\left(u'(r)\right)u(r);$$

$$P(r) = r^n \left[(p-1)\varphi\left(u'(r)\right)u'(r) + pF(r, u(r)) \right] \qquad (2.281)$$
$$+ (n-p)r^{n-1}\varphi\left(u'(r)\right)u(r),$$

where we again denote $F(r, u) = \int_0^u f(r, t)\, dt$. Observe that

$$P(0) = 0, \quad \text{and} \quad P(1) = \varphi\left(u'(1)\right)u'(1) > 0, \qquad (2.282)$$

provided that (2.265) holds, and

$$P'(r) = r^{n-1} \left[npF(r, u(r)) - (n-p)u(r)f(r, u(r)) + prF_r(r, u(r)) \right] (2.283)$$
$$\equiv r^{n-1}I(r).$$

We shall need the following lemma.

Lemma 2.44. *Let $u(r)$ be a positive solution of (2.264), and assume that the function $f(r, u)$ satisfies the condition (2.263). Then any solution $w(r)$ of the linearized problem (2.274) cannot vanish in the region where $f(r, u(r)) < 0$ (i.e., on $(r_2, 1)$, see (2.275)).*

Proof: Assume that the contrary is true, and let τ denote the largest root of $w(r)$ in the region where $f(r, u(r)) < 0$. We may assume that $w(r) > 0$ on $(\tau, 1)$. Then integrating the formula (2.279) over $(\tau, 1)$, we have

$$(p-1)\varphi\left(u'(1)\right)w'(1) - \tau^n(p-1)\varphi\left(u'(\tau)\right)w'(\tau)$$
$$= \int_\tau^1 \left(pf(r, u(r)) + rf_r(r, u(r)) \right)wr^{n-1}\, dr.$$

We have a contradiction, since the left hand side is non-negative, while the integral on the right is negative. $\diamond$

Next, we present the crucial non-degeneracy result.

Theorem 2.53. *Let $u(r)$ be a positive solution of (2.264). Assume that the conditions (2.263), (2.265), (2.267), (2.268), (2.272) and (2.273) hold. In the case $p < n$, assume additionally the following two conditions:*

$$u(r)f(r, u(r)) - pF(r, u(r)) > 0 \quad \text{for } r \in (0, r_2); \qquad (2.284)$$

$$I(r) \equiv npF(r, u(r)) - (n-p)u(r)f(r, u(r)) + prF_r(r, u(r)), \qquad (2.285)$$

defined in (2.283), satisfies any one of the following three conditions:

(i) $I(r) > 0$ *on* $(0, r_2)$,

(ii) $I(r) < 0$ *on* $(0, 1)$,

(iii) $I(r) > 0$ *on* $(0, r_0)$ *and* $I(r) < 0$ *on* $(r_0, 1)$, *for some* $r_0 \in (0, 1)$.

Then $u(r)$ is a non-singular solution, i.e., the corresponding linearized problem (2.274) admits only the trivial solution.

Proof: With r_2 as defined by (2.275), we claim that

$$Q(r) > 0 \quad \text{on} \ [0, r_2) \,. \tag{2.286}$$

In the case $p \geq n$, this follows immediately from the definition of $Q(r)$ in (2.280). In the other case, $p < n$, observe that $P(r) > 0$ on $[0, r_2)$ by (2.282) and (2.283). We write

$$Q(r) = P(r) + r^n \left[u(r) f(r, u(r)) - p F(r, u(r)) \right] ,$$

and hence $Q(r) > 0$ on $[0, r_2)$, by our condition (2.284).

We now define the function $O(r) = \gamma \xi(r) - T(r)$, which in view of (2.278) and (2.279) satisfies

$$O'(r) = \left[u(r) f_u(r, u(r)) - (p-1) f(r, u(r)) \right] w(r) r^{n-1} \left[\gamma - \alpha(r) \right] ,$$

with $\alpha(r)$ as defined by (2.273), and γ is a constant, to be selected. We may assume that $w(0) > 0$. We claim that the function $w(r)$ cannot have any roots inside $(0, 1)$. Assuming otherwise, let τ_1 be the smallest root of $w(r)$, i.e., $w(r) > 0$ on $[0, \tau_1)$. Let $\tau_2 \in (\tau_1, 1]$ denote the second root of $w(r)$. We now fix $\gamma = \alpha(\tau_1)$. By the monotonicity of $\alpha(r)$, the function $\gamma - \alpha(r)$ is non-positive on $[0, \tau_1)$, and non-negative on (τ_1, τ_2). Hence, $O'(r) \leq 0$ (at $r = \tau_1$, both $w(r)$ and $\gamma - \alpha(r)$ change sign). Since $O(0) = 0$, we conclude that

$$O(r) \leq 0, \quad \text{for all } r \in [0, \tau_2] \,. \tag{2.287}$$

There are two possibilities for the second root.

Case (i) $\tau_2 = 1$, i.e., $w(r) < 0$ on $(\tau_1, 1)$. From (2.287) we have $O(1) \leq 0$. On the other hand, $O(1) = -T(1) = -(p-1)\varphi\left(u'(1)\right) w'(1) > 0$, by (2.266), a contradiction.

Case (ii) $\tau_2 < 1$. Observe that $f(\tau_2, u(\tau_2)) > 0$. Indeed, assuming otherwise, we conclude by Lemma 2.42 that $f(r, u(r)) \leq 0$ on $(\tau_2, 1)$. But this contradicts Lemma 2.44. Applying Lemma 2.42 again, we conclude that $f(r, u(r)) > 0$ over $[0, \tau_2)$. It follows that $\tau_2 < r_2$, i.e., by (2.286), $Q(r) > 0$ on $[0, \tau_2)$.

Since $\xi(\tau_1) > 0$, while $\xi(\tau_2) < 0$, we can find a point $t \in (\tau_1, \tau_2)$, such that $\xi(t) = 0$, i.e.,

$$\frac{u(t)}{w(t)} = (p-1) \frac{\varphi(u'(t))}{\varphi'(u'(t)) \, w'(t)} = \frac{u'(t)}{w'(t)}. \tag{2.288}$$

Since $t \in (0, r_2)$, we have

$$Q(t) > 0. \tag{2.289}$$

In view of (2.287),

$$T(t) = -O(t) \geq 0.$$

On the other hand, using (2.288) and (2.289),

$$T(t) = \left[t^n \left((p-1)\varphi(u') \, w' \frac{u}{w} + f(t,u)u \right) + (n-p)t^{n-1}\varphi(u') \, u \right] \frac{w}{u}$$

$$= Q(t) \frac{w(t)}{u(t)} < 0,$$

giving us a contradiction.

It follows that $w(r)$ cannot have any roots, i.e., we may assume that $w(r) > 0$ on $[0, 1]$. But that is impossible, as can be seen by integrating (2.278) over $(0, 1)$. Hence $w \equiv 0$. $\diamond$

Our main example is the following model problem (here $r = |x|$)

$$\varphi(u'(r))' + \tfrac{n-1}{r}\varphi(u'(r)) - a(r)u^{p-1} + b(r)u^q = 0, \quad r \in (0, 1), \tag{2.290}$$

$$u = 0, \quad \text{for } r = 1.$$

We consider the sub-critical case

$$\min(1, p-1) < q < \frac{np - n + p}{n - p}, \tag{2.291}$$

and assume that the functions $a(r), b(r) \in C^1[0, 1]$ satisfy

$$a(r) > 0, \quad b(r) > 0, \quad a'(r) > 0, \quad b'(r) < 0 \quad \text{for } r \in (0, 1). \tag{2.292}$$

We define the functions

$$A(r) \equiv pa(r) + ra'(r),$$

$$B(r) \equiv \left(\frac{np}{q+1} - (n-p) \right) b(r) + \frac{prb'(r)}{q+1}.$$

Observe that $\frac{np}{q+1} - (n-p) > 0$ for subcritical q, i.e., when (2.291) holds.

Theorem 2.54. *In addition to the conditions (2.291) and (2.292), assume that the function $A(r)$ is positive and non-decreasing, while the function $B(r)$ is positive on $(0, 1)$. Assume also that the functions $\frac{rb'(r)}{b(r)}$ and $rb'(r)$ are non-increasing on $(0, 1)$. Then any positive solution of the problem (2.290) is non-singular.*

Proof: We shall verify the conditions of the Theorem 2.53. We have

$$uf - pF = b(r)u^{q+1}\left(1 - \frac{p}{q+1}\right) > 0 \quad \text{for all } r \in [0, 1),$$

verifying (2.284). Compute

$$(q - (p-1))\alpha(r) = -\frac{A(r)}{b(r)(u(r))^{q-p+1}} + p + \frac{rb'(r)}{b(r)}.$$

In view of our assumptions, $\alpha(r)$ is a non-increasing function, for all $r \in [0, 1)$, verifying the condition (2.273). Compute

$$I(r) = -A(r)u^p + B(r)u^{q+1} = B(r)u^p\left[-\frac{A(r)}{B(r)} + u^{q-p+1}\right].$$

Since by our conditions, $B(r)$ is non-increasing, it follows that the quantity in the square bracket is a non-increasing function, which is negative near $r = 1$. Hence, either $I(r)$ is negative over $(0, 1)$, or it changes sign exactly once, from positive to negative thus verifying either part (ii), or part (iii), of the condition (2.285). Hence, the Theorem 2.53 applies, implying that any positive solution of the problem (2.290) is non-singular.

Our second example is the problem

$$\varphi(u'(r))' + \frac{n-1}{r}\varphi(u'(r)) + b(r)u^q = 0, \quad r \in (0, 1) \tag{2.293}$$

$$u = 0, \quad \text{for } r = 1.$$

Similarly to the above, we establish the following result.

Theorem 2.55. *Let q satisfy (2.291), and assume that $b(r)$ satisfies*

$$b(r) > 0, \quad b'(r) < 0 \quad \text{for } r \in (0, 1),$$

the functions $\frac{rb'(r)}{b(r)}$ and $rb'(r)$ are non-increasing on $(0, 1)$.

Then any positive solution of the problem (2.290) is non-singular.

We consider next an autonomous problem $(f = f(u))$

$$\varphi(u'(r))' + \frac{n-1}{r}\varphi(u'(r)) + \lambda f(u) = 0, \quad r \in (0, 1), \tag{2.294}$$

$$u = 0 \text{ for } r = 1,$$

depending on a positive parameter λ. This problem is scaling invariant, which makes it easy to prove the following lemma, see B. Franchi, E. Lanconelli and J. Serrin [59]. (Recall that the condition (2.268) says that $u'(r) < 0$, so that $u(0)$ gives the maximum value of solution.)

Lemma 2.45. *Assume that the condition (2.268) holds. Then the maximum value of the positive solution of (2.294), $u(0) = \alpha$, uniquely identifies the solution pair $(\lambda, u(r))$ (i.e., there is at most one λ, with at most one solution $u(r)$, so that $u(0) = \alpha$).*

We now turn to the a priori estimates. Several people have already observed that the recent Liouville type result of J. Serrin and H. Zou [168], implies a priori estimates of B. Gidas and J. Spruck type [64], provided that the point of maximum of solution is bounded away from the boundary, see [11], [57]. In our case, positive solutions take their maximum at the origin, so that we have the following lemma, whose standard proof we sketch for completeness.

Lemma 2.46. *For the problem (2.262) assume that the conditions of Lemma 2.41 hold, and*

$$\lim_{u \to \infty} \frac{f(r, u)}{u^q} = b(r) \,,$$

with $\min(1, p-1) < q < \frac{np-n+p}{n-p}$, *and* $b(0) > 0$. *Then there exists a constant c, such any positive solution of the problem (2.262) satisfies*

$$|u|_{L^\infty} < c \,.$$

Proof: As we have already mentioned, the maximum of any positive solution occurs at $r = 0$, by Lemma 2.41. If there is a sequence of unbounded solutions, their rescaling tends to a non-trivial solution of

$$\Delta_p u + b(0)u^q = 0 \quad \text{in } R^n,$$

which is impossible by the result of J. Serrin and H. Zou [168], see [11] and [57] for more details.   ◇

We can now prove existence and uniqueness results for model equations.

Theorem 2.56. *Assuming the conditions of Theorem 2.55, the problem (2.293) has a unique positive solution.*

Proof: Assume first that $b(r) \equiv 1$. It is known that there exists a positive solution (proved by using variational methods). To see that the solution is unique, we consider a family of problems

$$\varphi(u'(r))' + \frac{n-1}{r}\varphi(u'(r)) + \lambda u^q = 0, \quad r \in (0, 1), \qquad (2.295)$$
$$u = 0, \quad \text{for } r = 1,$$

depending on a positive parameter λ. Since, by Theorem 2.55, positive solutions of (2.295) are non-singular, we can continue our solution at $\lambda = 1$, for both increasing and decreasing λ on a solution curve, in the Banach space X, defined on p. 382 of A. Aftalion and F. Pacella [11], and this solution curve does not admit any turns. (The space X, which is a subspace of $C^1(B)$, was originally defined in F. Pacard and T. Rivière [146]. A. Aftalion and F. Pacella [11] showed that one may apply the Implicit Function Theorem, when working in X.) By Lemma 2.40, solutions stay positive for all λ. By scaling, we see that the maximum value $u(0, \lambda) \to 0$, as $\lambda \to \infty$, and $u(0, \lambda) \to \infty$, as $\lambda \to 0$ (see Lemma 2.46). Since this solution curve covers all possible values of $u(0, \lambda)$, it follows by Lemma 2.45 that there is only one solution curve. We conclude existence and uniqueness of positive solution at all λ, in particular at $\lambda = 1$.

Turning to the general case, we consider a family of problems

$$\varphi(u'(r))' + \tfrac{n-1}{r}\varphi(u'(r)) + b^\theta(r)u^q = 0, \quad r \in (0,1),$$
$$u = 0 \text{ for } r = 1,$$

depending on a parameter $0 \le \theta \le 1$. It is easy to check that conditions of the Theorem 2.55 hold for all θ. Using the Theorem 2.55 and Lemma 2.46, we can continue the solutions between $\theta = 0$ and $\theta = 1$, to prove existence of positive solutions to the problem (2.293). Continuing backwards, between $\theta = 1$ and $\theta = 0$, we conclude the uniqueness of positive solution to the problem (2.293). $\diamondsuit$

Theorem 2.57. *Assuming the conditions of Theorem 2.54, the problem (2.290) has a unique positive solution.*

Proof: We consider a family of problems

$$\varphi(u'(r))' + \tfrac{n-1}{r}\varphi(u'(r)) - \theta a(r)u^{p-1} + b(r)u^q = 0, \quad r \in (0,1),$$
$$u = 0, \text{ for } r = 1,$$

depending on a parameter $0 \le \theta \le 1$. When $\theta = 0$, we have a unique positive solution, by the preceding result. Arguing as before, we conclude existence of positive solutions at $\theta = 1$. $\diamondsuit$

Chapter 3

Two Point Boundary Value Problems

3.1 Positive solutions of autonomous problems

We study global solution curves, and exact multiplicity of positive solutions of the Dirichlet problem

$$u''(x) + f(u(x)) = 0, \quad \text{for } -1 < x < 1, \ u(-1) = u(1) = 0. \tag{3.1}$$

Since positive solutions are symmetric with respect to the midpoint of the interval, it is convenient to pose the problem on the interval $(-1, 1)$. Our problem is autonomous, and so this does not restrict the generality. Any positive solution $u(x)$ is an even function, and $u'(x) < 0$ for $x \in (0, 1)$. This follows immediately by observing that any solution is symmetric with respect to any of its critical points (of course, the theorem of B. Gidas, W.-M. Ni and L. Nirenberg [63] is also applicable here). We have $u'(0) = 0$, and so this problem is just a one-dimensional version of equations on a unit ball, and so all of the results from the preceding chapter apply here. Of course, one-dimensional case allows for considerable simplifications and more detailed results. For example, the positivity of solutions for the linearized problem

$$w'' + f'(u(x))w = 0, \quad \text{for } -1 < x < 1, \ w(-1) = w(1) = 0, \tag{3.2}$$

one now obtains "for free", as the following lemma shows.

Lemma 3.1. *Assume $f(u) \in C^1(\bar{R}_+)$. Let $u(x)$ be a positive solution of (3.1), with*

$$u'(1) < 0. \tag{3.3}$$

If the problem (3.2) admits a nontrivial solution, then it does not change sign, i.e., we may assume that $w(x) > 0$ on $(-1, 1)$.

Proof: The function $u'(x)$ also satisfies the linear equation in (3.2). By the condition (3.3), $u'(x)$ is not a constant multiple of $w(x)$. Hence, by Sturm's Theorem, the roots of $u'(x)$ are interlaced with those of $w(x)$. If $w(x)$ had a root ξ inside say $(0,1)$, then $u'(x)$ would have to vanish on $(0,\xi)$, which is impossible. $\diamond$

We remark that the condition (3.3) holds, if $f(0) \geq 0$, as follows by Hopf's boundary lemma.

One can easily give conditions for solutions of (3.1) to be non-singular, i.e., for the problem (3.2) to have only the trivial solution. For example, if

$$f'(u) > \frac{f(u)}{u} \quad \text{(or if } f'(u) < \tfrac{f(u)}{u}\text{)}, \quad \text{for all } u > 0,$$

then the problem (3.2) has only the trivial solution. Indeed, writing the equation (3.1) in the form $u'' + \frac{f(u)}{u} u = 0$, we conclude by Sturm's comparison theorem that there is a root of $w(x)$ between the roots ± 1 of $u(x)$, which is impossible, since $w(x)$ is positive. In case $f(u)$ changes sign from negative to positive, one can do much better, as the following theorem shows.

Theorem 3.1. *Assume* $f(u) \in C^1(\bar{R}_+)$, *and* $f(u) < 0$ *on* $(0,\gamma)$, *while* $f(u) > 0$ *on* (γ, ∞), *for some* $\gamma > 0$. *Assume that*

$$f'(u) > \frac{f(u)}{u}, \quad \text{for all } u > \gamma. \tag{3.4}$$

Then any positive solution of (3.1), satisfying (3.3), is non-singular, i.e., the problem (3.2) admits only the trivial solution.

Proof: Assume, on the contrary, that the problem (3.2) admits a non-trivial solution $w(x) > 0$. Let $x_0 \in (0,1)$ denote the point where $u(x_0) = \gamma$, i.e., the condition (3.4) holds for $0 \leq x < x_0$. From the equations (3.1) and (3.2), we get

$$(w'u - wu')' = -\left[f'(u) - \frac{f(u)}{u}\right] uw < 0, \quad \text{for } 0 \leq x < x_0.$$

Integrating this over $(0, x_0)$,

$$w'(x_0)u(x_0) - w(x_0)u'(x_0) < 0. \tag{3.5}$$

We claim that

$$(x_0 - 1)u'(x_0) - u(x_0) > 0. \tag{3.6}$$

Indeed, letting $q(x) \equiv (x-1)u'(x) - u(x)$, we see that $q(1) = 0$, while $q'(x) = (x-1)u''(x) < 0$ on $(x_0, 1)$, justifying (3.6). Next, we observe that the function $u''(x)w(x) - u'(x)w'(x)$ is constant over $[0, 1]$. (Just differentiate this function, and use the corresponding equations.) Evaluating this function at $x = x_0$, and at $x = 1$, and observing that $u''(x_0) = -f(u(x_0)) = -f(\gamma) = 0$, we have

$$-u'(x_0)w'(x_0) = -u'(1)w'(1) \,, \tag{3.7}$$

which implies, in particular, that

$$w'(x_0) < 0 \,. \tag{3.8}$$

The function $z(x) \equiv xu'(x)$ is easily seen to satisfy

$$z'' + f'(u)z = -2f \,.$$

Using the corresponding equations, we express

$$(z'w - w'z)' = -2fw > 0 \,, \quad \text{for } x \in (x_0, 1) \,.$$

Integrating this over $(x_0, 1)$, we get (observe that $z' = u' + xu''$, $z'(x_0) = u'(x_0)$)

$$L \equiv -u'(1)w'(1) - u'(x_0)w(x_0) + x_0 u'(x_0)w'(x_0) > 0 \,.$$

On the other hand, using (3.7) and (3.5), and then (3.6) and (3.8), we have

$$L < -u'(x_0)w'(x_0) - w'(x_0)u(x_0) + x_0 u'(x_0)w'(x_0) \tag{3.9}$$
$$= w'(x_0) \left[(x_0 - 1)u'(x_0) - u(x_0) \right] < 0 \,,$$

a contradiction. $\diamond$

It is remarkable that no assumptions on $f(u)$ are required in the region, where it is negative. This result easily implies uniqueness of positive solutions to (3.1). Uniqueness part was proved previously by J. Cheng [35], using time maps.

We shall discuss positive solutions in case $f(u)$ is a cubic, with three distinct roots

$$u''(x) + \lambda(u-a)(u-b)(c-u) = 0, \quad \text{for } -1 < x < 1, \tag{3.10}$$
$$u(-1) = u(1) = 0 \,.$$

The most interesting case is when $0 \le a < b < c$. Indeed, if we have, say, $a < 0 < b < c$, then we have a convex-concave nonlinearity with $f(0) = abc < 0$. It follows, by the Theorem 2.7, that there is at most one turn on any solution curve. Using the methods of the preceding chapter, it is easy to show that if the area of the positive hump of the cubic exceeds that of the negative hump, there is exactly one parabola-like curve, with a turn to the right. The case when $0 = a < b < c$, was already covered above, even in the more general case of balls in R^n (observe that we may replace the boundary conditions in (3.10) by $u'(0) = u(1) = 0$.) So, we shall assume that all three roots are positive. This problem was studied in a 1981 paper by J. Smoller and A. Wasserman [174]. They had succeeded in proving an exact multiplicity result in case $a = 0$, while their proof for $a > 0$ case contained an error. This error was discovered by S.-H. Wang [188], who was able to fix the proof, but under some restriction on $a > 0$. Both of these papers used time maps. Then, P. Korman, Y. Li and T. Ouyang [108] used bifurcation theory to attack the problem, but again some restriction on $a > 0$ was necessary. (All of the above mentioned papers covered more general $f(u)$, behaving like a cubic.) Finally, P. Korman, Y. Li and T. Ouyang [111], building on their previous work, had given a computer assisted proof of exact multiplicity result for the general cubic. It turns out that the set of all positive solutions of (3.10) consists of two curves, with the lower curve monotone in λ, and the upper curve having exactly one turn. The numerical computations in P. Korman, Y. Li and T. Ouyang [111] also showed that the approach in J. Smoller and A. Wasserman [174] could not possibly cover the general cubic. (Their approach required a certain integral to be positive, in order to derive a differential inequality for the time map. However, computations show that the integral changes sign for some choices of a, b and c.)

The maximum value of positive solution is $\alpha = u(0)$. Clearly, either $\alpha \in (0, a)$, or $\alpha \in (b, c)$, so that $f(\alpha) > 0$. A necessary condition for existence of positive solutions, with $\alpha \in (b, c)$, is

$$\int_a^c (u - a)(u - b)(c - u)\, du > 0, \tag{3.11}$$

i.e., the area of the positive hump in the graph of the cubic exceeds that of the negative hump. Indeed, if this condition was violated, we would obtain a contradiction, multiplying the equation (3.10) by $u'(x)$, and integrating between $x = 0$ and the point ξ, where $u(\xi) = a$. The integral in (3.11) evaluates to $\frac{1}{12}(c - a)^3(a - 2b + c)$, and it is positive if

$$c > 2b - a. \tag{3.12}$$

The optimal computer assisted result of P. Korman, Y. Li and T. Ouyang [111] is the following (see also P. Korman [92], for an exposition).

Theorem 3.2. *Assume that the necessary condition (3.12) holds. Then there exists a critical λ_0, such that the problem (3.10) has exactly one positive solution for $0 < \lambda < \lambda_0$, exactly two positive solutions at $\lambda = \lambda_0$, and exactly three positive solutions for $\lambda_0 < \lambda < \infty$. Moreover, all solutions lie on two smooth solution curves. One of the curves, referred to as the lower curve, starts at $(\lambda = 0, u = 0)$, it is increasing in λ, and $\lim_{\lambda \to \infty} u(x, \lambda) = a$ for $x \in (-1, 1)$. The upper curve is a parabola-like curve, with exactly one turn to the right. (See Figure 3.1.)*

In the bifurcation diagram we use $u(0)$, i.e., the L^∞ norm, to identify the solution.

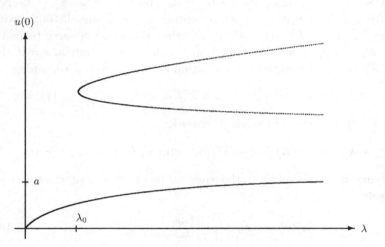

Fig. 3.1   Bifurcation diagram for a cubic

3.2   Direction of the turn

We consider again positive solutions for the problem

$$u'' + \lambda f(u) = 0, \quad -1 < x < 1, \quad u(-1) = u(1) = 0. \tag{3.13}$$

The corresponding linearized problem is

$$w'' + \lambda f'(u)w = 0, \quad -1 < x < 1, \quad w(-1) = w(1) = 0. \tag{3.14}$$

If $u(x)$ is a positive solution of (3.13), and $f(0) \geq 0$, then in view of Hopf's boundary lemma

$$u'(1) < 0, \tag{3.15}$$

and the Crandall-Rabinowitz Theorem 1.2 applies (as we saw in the preceding chapter). In case $f(0) < 0$, we shall assume that the condition (3.15) holds at any singular solution. Recall that a solution $(\lambda_0, u_0(x) > 0)$ of (3.13) is called singular if the corresponding linearized problem (3.14) admits a non-trivial solution. Let us recall how to compute the direction of bifurcation at a singular point $(\lambda_0, u_0(x))$. According to the Crandall-Rabinowitz Theorem 1.2, the solution set near the point $(\lambda_0, u_0(x))$ is a curve $\lambda = \lambda(s)$, $u = u(s)$, with $\lambda(0) = \lambda_0$ and $u(0) = u_0(x)$. Observe that $\lambda'(0) = 0$, and $u_s(0) = w(x)$, according to the Crandall-Rabinowitz Theorem 1.2. If we show that $\lambda''(0) < 0$, then the solution curve turns to the left at $(\lambda_0, u_0(x))$, since the function $\lambda(s)$ has a maximum at $s = 0$. To express $\lambda''(s)$, we differentiate the equation (3.13) twice in s, obtaining

$$u''_{ss} + \lambda f_u u_{ss} + \lambda f_{uu} u_s^2 + 2\lambda' f_u u_s + \lambda'' f = 0, \quad u_{ss}(-1) = u_{ss}(1) = 0.$$

Letting $s = 0$, we have by the above remarks

$$u''_{ss} + \lambda_0 f_u u_{ss} + \lambda_0 f_{uu} w^2 + \lambda''(0) f = 0, \quad u_{ss}(-1) = u_{ss}(1) = 0.$$

Multiplying this equation by w, the equation (3.14) by u_{ss}, subtracting and integrating

$$\lambda''(0) = -\lambda_0 \frac{\int_{-1}^{1} f_{uu}(\lambda_0, u_0(x)) w^3(x)\, dx}{\int_{-1}^{1} f(\lambda_0, u_0) w(x)\, dx} \tag{3.16}$$

$$= -\lambda_0 \frac{\int_0^1 f_{uu}(\lambda_0, u_0(x)) w^3(x)\, dx}{\int_0^1 f(\lambda_0, u_0) w(x)\, dx},$$

where at the last step we used the fact that $u_0(x)$ and $w(x)$ are both even functions.

The following result on the direction of the turn we used in Chapter 2 for balls in R^n. We present next a simple proof for the $n = 1$ case.

Theorem 3.3. **(i)** *Assume $f(0) \leq 0$, and $f(u)$ is convex-concave. Then at any singular point, satisfying $u'_0(1) < 0$, we have $\lambda''(0) > 0$, and hence a turn to the right occurs.*
(ii) *Assume $f(0) \geq 0$, and $f(u)$ is concave-convex. Then at any singular point we have $\lambda''(0) < 0$, and hence a turn to the left occurs.*

Proof: To simplify the notation, we shall write (λ, u) instead of (λ_0, u_0). By Lemma 3.1, $w(x) > 0$ on $(-1, 1)$, which implies that

$$w'(1) < 0 .$$

From the equations (3.13) and (3.14) it is straightforward to verify the following identities

$$u'(x)w'(x) - u''(x)w(x) = constant = u'(1)w'(1) ; \tag{3.17}$$

$$(u''w' - u'w'')' = \lambda f''(u)u'^2 w . \tag{3.18}$$

Assume that the first set of conditions hold. Integrating (3.18),

$$\lambda \int_0^1 f''(u)u'^2 w \, dx = u''(1)w'(1) = -\lambda f(0)w'(1) \leq 0 . \tag{3.19}$$

Consider the function $p(x) \equiv \frac{w(x)}{-u'(x)}$. Since by (3.17)

$$p'(x) = -\frac{u'(1)w'(1)}{u'(x)^2} < 0 ,$$

the function $p(x)$ is positive and decreasing on $(0, 1)$. The same is true for $p^2(x) = \frac{w^2(x)}{u'^2(x)}$. Let x_0 be the point where $f''(u(x))$ changes sign (i.e., $f''(u(x)) < 0$ on $(0, x_0)$, and $f''(u(x)) > 0$ on $(x_0, 1)$). By scaling $w(x)$, we may achieve that $w^2(x_0) = u'^2(x_0)$. Then $w^2(x) > u'^2(x)$ on $(0, x_0)$, and the inequality is reversed on $(x_0, 1)$. Using (3.19), we have

$$\int_0^1 f''(u)w^3 \, dx < \int_0^1 f''(u)u'^2 w \, dx \leq 0 . \tag{3.20}$$

Integrating (3.17), we have

$$\int_0^1 f(u)w \, dx = \frac{1}{2\lambda}u'(1)w'(1) > 0 . \tag{3.21}$$

The formulas (3.20) and (3.21), together with (3.16), imply that $\lambda''(0) > 0$. The second part of the theorem is proved similarly. $\diamondsuit$

3.3 Stability and instability of solutions

We consider the question of stability and instability of positive solutions for the problem

$$u'' + \lambda f(u) = 0, \quad -1 < x < 1, \quad u(-1) = u(1) = 0 . \tag{3.22}$$

Recall that the maximum value of solution, $\alpha = u(0)$, uniquely identifies the solution pair $(\lambda, u(x))$, by Lemma 2.2. Hence, the solution set of (3.22) can be faithfully depicted by planar curves in (λ, α) plane. It is natural to ask: which way the solution curve travels through a given point (λ, α)? One approach, using time maps, has been known for a while, see K.J. Brown et al [26]. Namely, denoting $F(u) = \int_0^u f(t)\,dt$, and $h(u) = 2F(u) - uf(u)$, one has

$$\frac{d}{d\alpha}\lambda(\alpha)^{1/2} = \frac{1}{\sqrt{2}} \int_0^1 \frac{h(\alpha) - h(\alpha v)}{[F(\alpha) - F(\alpha v)]^{3/2}}\,dv.$$

We see that $\frac{d\lambda}{d\alpha} < 0$ (> 0), and the curve travels to the left (right) in (λ, α) plane, provided that $h(\alpha) < h(u)$ $(h(\alpha) > h(u))$ for all $u \in (0, \alpha)$. Time maps work only for autonomous ODE's. It was desirable to find a more flexible approach, which could be applied to other situations. One way to avoid using time maps was given by P. Korman and J. Shi [117]. Then, P. Korman [94] found another way, which is both simpler, and it gives more general results. We shall follow the approach in P. Korman [94].

The eigenvalue problem for the linearized equation in (3.22) is

$$w'' + \lambda f'(u)w + \mu w = 0, \quad -1 < x < 1, \quad w(-1) = w(1) = 0. \qquad (3.23)$$

We shall be particularly interested in the principal (the smallest) eigenvalue, which we denote by μ_1. The corresponding eigenfunction, which we shall call $w(x)$, is of one sign, i.e., we may assume that $w(x) > 0$ on $(-1, 1)$. Recall that any positive solution of (3.22) is an even function, which satisfies $u'(x) < 0$ $(u'(x) > 0)$ for $x > 0$ $(x < 0)$, and hence $\alpha = u(0)$ is its maximum value. Observe that the principal eigenfunction $w(x)$ is also an even function. Indeed, assuming otherwise, $w(-x)$ would give us a linearly independent solution to the problem (3.23), contradicting the simplicity of the eigenvalues of (3.23). Recall that the solution $u(x)$ of (3.22) is called *stable* (*unstable*), if $\mu_1 > 0$ $(\mu_1 < 0)$. (In case $\mu_1 = 0$, $u(x)$ is sometimes called *neutrally stable*.) Define

$$h(u) = 2F(u) - uf(u),$$

where, as usual, $F(u) = \int_0^u f(t)\,dt$.

Theorem 3.4. *Assume that*

$$h(\alpha) < h(u), \quad \text{for all } u < \alpha. \qquad (3.24)$$

Then the positive solution of (3.22), with maximum value $u(0) = \alpha$, is unstable i.e., $\mu_1 < 0$.

Proof: Assume, on the contrary, that $\mu_1 \geq 0$. We claim that

$$u'(x)w'(x) - u''(x)w(x) > 0, \quad \text{for all } x \in (-1, 1). \tag{3.25}$$

Indeed, denoting $p(x) \equiv u'(x)w'(x) - u''(x)w(x)$, we see that $p(0) = -w(0)u''(0) > 0$, and

$$p'(x) = -\mu_1 w(x)u'(x). \tag{3.26}$$

Since $p(x)$ is increasing (decreasing) for $x > 0$ ($x < 0$), the claim follows.

From the equations (3.22) and (3.23), we have

$$\int_{-1}^{1} h'(u(x))w(x)\,dx = \int_{-1}^{1} (f(u) - uf'(u))\,w\,dx = \frac{\mu_1}{\lambda} \int_{-1}^{1} uw\,dx \geq 0. \tag{3.27}$$

On the other hand, in view of (3.24) and (3.25),

$$\int_{-1}^{1} h'(u(x))w(x)\,dx = \int_{-1}^{1} \frac{d}{dx}\left[h(u) - h(\alpha)\right]\frac{w}{u'}\,dx \tag{3.28}$$

$$= -\int_{-1}^{1} \left[h(u) - h(\alpha)\right]\frac{u'(x)w'(x) - u''(x)w(x)}{u'^2}\,dx < 0,$$

which is a contradiction. Observe that the last integral in (3.28) is proper, since $h(u(x)) - h(\alpha)$ is asymptotically quadratic in x, near $x = 0$, and the same is true for the denominator, $u'^2(x)$. $\Diamond$

The trick of introducing the $h(u) - h(\alpha)$ term was inspired by R. Schaaf and K. Schmitt [162]. Similarly, we have a stability result.

Theorem 3.5. *Assume that*

$$h(\alpha) > h(u), \quad \text{for all } u < \alpha. \tag{3.29}$$

Then the positive solution of (3.22), with maximum value $u(0) = \alpha$, is stable.

Proof: Assume, on the contrary, that $\mu_1 \leq 0$. Again, we claim that $p(x) = u'(x)w'(x) - u''(x)w(x)$ is positive on $(-1, 1)$. Indeed, $p(\pm 1) = u'(\pm 1)w'(\pm 1) > 0$, while (3.26) implies that $p(x)$ is decreasing (increasing) for $x > 0$ ($x < 0$), and the claim follows. The rest of the proof is the same as in the Theorem 3.4. The inequality signs are now reversed, in both of the inequalities (3.27) and (3.28), and again we reach a contradiction. $\Diamond$

For the problem (3.22) it is known that, when $\alpha = u(0)$ is increasing, the solution curve travels northeast in the (λ, α) plane, if and only if the solution is stable, and northwest, if and only if it is unstable. This can be deduced, for example, from the Proposition 4.1.3 in R. Schaaf [161]. We give another proof, although we use some ideas from [161]. We shall need the following version of the Sturm comparison theorem.

Lemma 3.2. *Assume that the functions $v(x)$ and $w(x)$, of class C^2, satisfy*

$$v'' + a(x)v \geq 0, \quad v'(0) = 0, \quad v(0) > 0,$$

$$w'' + a(x)w \leq 0, \quad w'(0) = 0, \quad w(0) > 0,$$

on some interval $(0, \gamma)$, with a given continuous function $a(x)$, and with at least one of the inequalities being strict on a set of positive measure. Then $w(x)$ oscillates faster than $v(x)$. Namely, if $w(x) > 0$ on $(0, \gamma)$, then $v(x) > 0$ on $(0, \gamma)$. If, on the other hand, $v(\gamma) = 0$, then $w(x)$ must vanish inside $(0, \gamma)$.

Proof: Assume that $w(x) > 0$ on $(0, \gamma)$, but $v(\xi) = 0$ for some $\xi \in (0, \gamma)$, while $v(x) > 0$ on $(0, \xi)$. On $(0, \xi)$ we have

$$(v'w - w'v)' \geq 0,$$

and the inequality is strict on a set of positive measure. Integrating this over $(0, \xi)$, we obtain a contradiction. The other statement is proved similarly.

◇

Proposition 3.1. *Let $u(x, \alpha)$ denote a positive solution of (3.22), with $u(0, \alpha) = \alpha$. Assume that $u_x(1, \alpha) < 0$. Then $\mu_1 < 0$ $(\mu_1 > 0)$ if and only if $\lambda'(\alpha) < 0$ $(\lambda'(\alpha) > 0)$.*

Proof: By shifting and stretching of the interval, we can convert the problem (3.22) into

$$u'' + f(u) = 0, \quad -L < x < L, \quad u(-L) = u(L) = 0, \qquad (3.30)$$

with $L = \sqrt{\lambda}$. We use "shooting", with $u(0) = \alpha$ and $u'(0) = 0$, and we have $L = L(\alpha)$, i.e., $L(\alpha) = \sqrt{\lambda(\alpha)}$. We shall calculate $L'(\alpha)$, which has the same sign as $\lambda'(\alpha)$. Differentiating the relation $u(L(\alpha), \alpha) = 0$, we have

$$u_x(L(\alpha), \alpha)L'(\alpha) + u_\alpha(L(\alpha), \alpha) = 0.$$

Since by our assumption $u_x(L(\alpha), \alpha) < 0$, it follows that the sign of $L'(\alpha)$ is the same as that of $u_\alpha(L(\alpha), \alpha)$. To determine the latter, we differentiate the equation (3.30) in α

$$u_\alpha'' + f'(u)u_\alpha = 0, \quad u_\alpha'(0, \alpha) = 0, \quad u_\alpha(0, \alpha) = 1.$$

If $\mu_1 > 0$, then

$$w'' + f'(u)w = -\mu_1 w < 0,$$

and hence by Lemma 3.2, w oscillates faster than u_α, and then $u_\alpha(L(\alpha), \alpha) > 0$. In case $\mu_1 < 0$, u_α has to vanish at some $\theta \in (0, L)$, since it oscillates faster than w. We claim that $u_\alpha(L(\alpha), \alpha) < 0$. Indeed, the function $u'(x)$ satisfies the same linear equation as u_α, and hence their roots interlace. But $u'(x) < 0$ on $(0, L)$, and so u_α cannot have any more roots, in addition to θ.

It is well known that $L'(\alpha) = 0$, if and only if $\mu_1 = 0$, see e.g., [108]. Hence, conversely, $L'(\alpha) > 0 \, (< 0)$ implies that $\mu_1 > 0 \, (< 0)$. $\diamondsuit$

In the following example we pose the problem on the interval $(0, 1)$, to make it consistent with our *Mathematica* program.

Example The curve of positive solutions (in the (λ, α) plane, with $\alpha = umax = u(1/2)$) for the problem

$$u'' + \lambda u(2 + \sin u) = 0, \quad 0 < x < 1, \quad u(0) = u(1) = 0 \tag{3.31}$$

has infinitely many turns (see the Figure 3.2, for the bifurcation diagram computed using *Mathematica*).

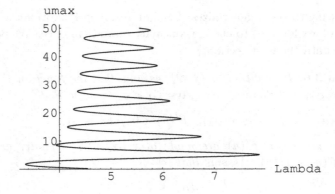

Fig. 3.2   The bifurcation diagram for the problem (3.31)

Indeed, by the Crandall-Rabinowitz Theorem [39], there is a curve of positive solutions bifurcating from zero. As we follow this curve for increasing $\alpha = u(1/2)$, we shall use the Theorems 3.4 and 3.5 to show that there are infinitely many changes of stability, which implies that there are infinitely many changes in the direction of the curve, by the Proposition 3.1 above. Indeed, here $h(u) = -u^2 \sin u - 2u \cos u + 2 \sin u$. This function oscillates about zero, with increasing amplitude. Clearly, there is a sequence $\{\alpha_n\} \to \infty$, so that $h(u) > h(\alpha_n)$ for all $u \in (0, \alpha_n)$, and another sequence $\{\beta_n\} \to \infty$, so that $h(u) < h(\beta_n)$ for all $u \in (0, \beta_n)$. Solutions with $umax = u(1/2) = \alpha_n$ are unstable, and the ones with $u(1/2) = \beta_n$ are stable. We claim that along the solution curve, λ cannot go to either zero or infinity. To see that λ cannot go to zero, just multiply the equation (3.31) by u, and integrate. Using Poincare's inequality, we get a contradiction if $\lambda \to 0$. If we had $\lambda \to \infty$ along the solution curve, then by Sturm's comparison theorem the solutions would have to become sign-changing, contradicting the fact that they are positive. It follows that the solution curve is constrained to a bounded in λ strip, and it makes infinitely many turns.

3.3.1   *S-shaped curves of combustion theory*

We consider the problem (3.22), and assume that $f(u) \in C^2[0, \bar{u}]$ for some $\bar{u} \leq \infty$, and it satisfies

$$f(u) > 0, \quad \text{for } u \in [0, \bar{u}). \tag{3.32}$$

The following theorem generalizes the main result in P. Korman and Y. Li [103]. It allows for $f(u)$ to change concavity more than once. We postpone the proof until the next section.

Theorem 3.6. *In addition to (3.32), assume that the function $f''(u)$ has a root $u = \alpha_1$, and there is an α_2, $0 < \alpha_1 < \alpha_2 < \bar{u}$, so that*

$$f''(u) > 0 \ \text{for } u \in (0, \alpha_1), \ \ f''(u) < 0 \ \text{for } u \in (\alpha_1, \alpha_2). \tag{3.33}$$

(No assumptions on $f''(u)$ are made over $(\alpha_2, \bar{u})$.) Denote, as before, $h(u) = 2F(u) - uf(u)$. Assume that

$$h(\alpha_1) < 0, \tag{3.34}$$

$$h(\alpha_2) > h(u), \quad \text{for all } u \in (0, \alpha_2), \tag{3.35}$$

$$h'(u) > 0, \quad \text{for all } u \in (\alpha_2, \bar{u}). \tag{3.36}$$

Then the solution curve for (3.22) is exactly S-shaped (i.e., it begins at $(\lambda, \alpha) = (0, 0)$, and it makes exactly two turns).

Remark If $\bar{u} < \infty$, and $f(\bar{u}) = 0$, then, by the results of the previous chapter, the curve exist for all $\lambda > 0$, and it tends to $\bar{u}$ as $\lambda \to \infty$. If $\bar{u} = \infty$, then integrating the inequality $h'(u) \geq 0$, we see that $f(u)$ is below a linear function for large u. If $f(u)$ is asymptotically linear, then the solution curve will go to infinity at some finite λ. If $\lim_{u \to \infty} \frac{f(u)}{u} = 0$, then the solution curve continues for all $\lambda > 0$, see P. Korman and Y. Li [103].

The Theorem 3.6 is applicable to the original equations of combustion theory (see R. Aris [15])

$$v'' + \mu(1 - \epsilon v)^m e^{\frac{v}{1+\epsilon v}} = 0, \quad x \in (-1,1), \quad v(-1) = v(1) = 0, \qquad (3.37)$$

where μ, m and ϵ are positive parameters. (If $m = 0$, we have the so called perturbed Gelfand problem.) For sufficiently small ϵ, S.P. Hastings and J.B. McLeod [68] proved that the solution curve for (3.37) is exactly S-shaped. This result was improved and extended by S.-H. Wang and F.P. Lee [191], and S.-H. Wang [190]. All of these papers used a rather complicated time map (or quadrature) method. We shall use our results above.

In our experiments, when $\epsilon = \epsilon(m)$ was small, the Theorem 3.6 applied, giving a S-shaped curve, while for larger ϵ the solution curve is monotone. Before we give the examples, it is convenient to rescale $\epsilon v = u$, and $\lambda = \mu \epsilon$, and consider

$$u'' + \lambda(1 - u)^m e^{\frac{u}{\epsilon(1+u)}} = 0, \quad x \in (-1,1), \quad u(-1) = u(1) = 0. \qquad (3.38)$$

We are interested in the solutions satisfying $0 < u(x) < 1$, for all $x \in (-1,1)$. The following examples are also covered by the results of [191] and [190], although our approach appears to be simpler.

Example 1 Let $m = 2$, and $\epsilon = 0.05$, i.e., we consider the problem

$$u'' + \lambda(1 - u)^2 e^{\frac{u}{0.05(1+u)}} = 0, \quad x \in (-1,1), \quad u(-1) = u(1) = 0. \qquad (3.39)$$

Here $\bar{u} = 1$. With the help of *Mathematica*, one verifies that the second derivative $f''(u)$ goes plus-minus-plus on the interval $(0,1)$, with the roots $\alpha_1 \simeq 0.52$ and $\alpha_2 \simeq 0.9$. Also $h(\alpha_1) \simeq -49.6 < 0$. One also verifies that $h(\alpha_2) > h(u)$, for all $u \in (0, \alpha_2)$, and $h'(u) > 0$ for $u > \alpha_2$. Hence, the Theorem 3.6 applies, and the solution curve of (3.39) is exactly S-shaped (see the curve on the left in the Figure 3.3).

Example 2 Let $m = 2$, and $\epsilon = 0.2$, i.e., we consider the problem

$$u'' + \lambda(1 - u)^2 e^{\frac{u}{0.2(1+u)}} = 0, \quad x \in (-1,1), \quad u(-1) = u(1) = 0. \qquad (3.40)$$

This time, $f(0) > 0$, and $f(u)$ changes concavity once, at $\alpha \simeq 0.7$, with $f''(u) < 0$ for $u \in (0, \alpha)$ and $f''(u) > 0$ for $u \in (\alpha, 1)$. According to the

Theorem 2.8, the solution curve, which starts at ($\lambda = 0, u = 0$), can make only turns to the left. But if such a turn had occurred, the solution curve would have nowhere to go, since no more turns are possible, while solutions are bounded by 1. Hence, the solution curve is monotone in λ. In other words, we have existence and uniqueness of positive solution for all $\lambda > 0$ (see the curve on the right in the Figure 3.3).

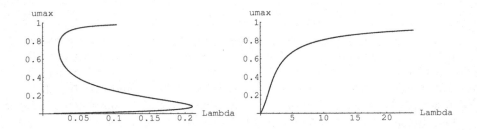

Fig. 3.3   The bifurcation diagrams for $\epsilon = 0.05$, and $\epsilon = 0.2$

3.3.2   *An extension of the stability condition*

One can generalize the stability result, the Theorem 3.5, in case $f(u) > 0$. With $h(u) = 2F(u) - uf(u)$ as before, we define $g(u) = \int_0^u h(t)\,dt$. The following result was proved in P. Korman [94].

Theorem 3.7. *Assume that $f(u) > 0$ for all $u > 0$, and*

$$g(u) > g(\alpha) + g'(\alpha)(u - \alpha), \quad \text{for all } u \in [0, \alpha). \tag{3.41}$$

Then any positive solution of (3.22), with $u(0) = \alpha$, is stable.

It is easy to see that the condition (3.41) is more general than (3.29). Indeed, denoting $G(u) = g(u) - g(\alpha) - g'(\alpha)(u - \alpha)$, we see that $G(\alpha) = 0$, and $G'(u) = h(\alpha) - h(u) < 0$, provided that condition (3.29) holds. But then $G(u) > 0$ for $u \in [0, \alpha)$, i.e., (3.41) holds.

3.4 *S*-shaped solution curves

Following P. Korman and Y. Li [103], we now consider a case when the curve of positive solutions of

$$u'' + \lambda f(u) = 0, \quad \text{for } -1 < x < 1, \ u(-1) = u(1) = 0 \tag{3.42}$$

is exactly *S*-shaped. The main application will be to the "perturbed Gelfand equation" of combustion theory. We shall need the corresponding linearized problem

$$w'' + \lambda f'(u)w = 0, \quad -1 < x < 1, \ w(-1) = w(1) = 0. \tag{3.43}$$

Recall that $(\lambda, u(x))$ is called a *critical point* (or a singular solution) of (3.42), if the problem (3.43) admits non-trivial solutions. We assume that $f(u) \in C^2[0, \bar{u}]$ for some $0 < \bar{u} \leq \infty$, and that it satisfies

$$f(u) > 0 \quad \text{for all } 0 \leq u < \bar{u}. \tag{3.44}$$

(Observe that here $f(0) > 0$.) We assume $f(u)$ to be convex-concave, i.e., there is an $\alpha \in (0, \bar{u})$, such that

$$f''(u) > 0 \text{ for } u \in (0, \alpha), \ f''(u) < 0 \text{ for } u \in (\alpha, \bar{u}). \tag{3.45}$$

We define a function $I(u) \equiv f^2(u) - 2F(u)f'(u)$, where $F(u) = \int_0^u f(t)\,dt$. We assume that there is a $\beta > \alpha$, such that

$$I(\beta) = f^2(\beta) - 2F(\beta)f'(\beta) \geq 0. \tag{3.46}$$

The following lemma has originated from P. Korman, Y. Li and T. Ouyang [108], and P. Korman and Y. Li [103]. (Recall that both $u(x)$ and $w(x)$ are even functions, so that we may restrict the integrals below to the interval $(0, 1)$.)

Lemma 3.3. *Assume that $f(u)$ satisfies the conditions (3.44)-(3.46). Let (λ, u) be any critical point of (3.42), such that $u(0) \geq \beta$, and let $w(x)$ be any non-trivial solution of the linearized problem (3.43). Then*

$$\int_0^1 f''(u(x))u'(x)w^2(x)\,dx > 0. \tag{3.47}$$

Proof: We shall derive a convenient expression for the integral in (3.47). Differentiating (3.43), we get

$$w_x'' + \lambda f'(u)w_x + \lambda f''(u)u'w = 0. \tag{3.48}$$

Multiplying the equation (3.48) by w, the equation (3.43) by w_x, integrating over $(0, 1)$ and subtracting, we express

$$\lambda \int_0^1 f''(u) u' w^2 \, dx = w'^2(1) - \lambda w^2(0) f'(u(0)). \tag{3.49}$$

By differentiation, we verify that $u''(x)w(x) - u'(x)w'(x)$ is equal to a constant for all x, and hence

$$u''(x)w(x) - u'(x)w'(x) = -\lambda w(0) f(u(0)), \quad \text{for all } x \in [0, 1]. \tag{3.50}$$

Evaluating this expression at $x = 1$, we obtain

$$w'(1) = \frac{\lambda w(0) f(u(0))}{u'(1)}. \tag{3.51}$$

Multiplying (3.42) by u', and integrating over $(0, 1)$, we have

$$u'^2(1) = 2\lambda F(u(0)). \tag{3.52}$$

Using (3.52) and (3.51) in (3.49), we finally express

$$\int_0^1 f''(u) u' w^2 \, dx = \frac{w^2(0)}{2F(u(0))} I(u(0)). \tag{3.53}$$

By our assumption (3.46), $I(\beta) \geq 0$. Since

$$I'(u(0)) = -2F(u(0)) f''(u(0)) > 0 \quad \text{for } u(0) \geq \beta,$$

we conclude that $I(u(0)) > I(\beta) \geq 0$, and, in view of (3.53), the lemma follows. $\diamondsuit$

The following lemma contains the crucial trick, which has originated from P. Korman, Y. Li and T. Ouyang [108]. It says that for convex-concave problems, with $f(0) > 0$, only turns to the right are possible in the (λ, α) plane, once the maximum value of the solution, $u(0)$, has reached a certain level. In case $f(0) \leq 0$, a similar result was proved in the preceding chapter, holding even for the PDE case.

Lemma 3.4. *In the conditions of the preceding Lemma 3.3, assume again that $u(0) \geq \beta$, and $w(x) > 0$ is a solution of the corresponding linearized problem (3.43). Then*

$$\int_0^1 f''(u(x)) w^3(x) \, dx < 0. \tag{3.54}$$

Proof: Let $(\lambda, u(x))$ be a critical point of (3.42). Since $u(0) \geq \beta > \alpha$, it follows that the function $f''(u(x))$ changes sign exactly once on $(0,1)$, say at x_0. Then we have

$$f''(u(x)) < 0 \text{ for } x \in (0, x_0), \quad f''(u(x)) > 0 \text{ for } x \in (x_0, 1). \qquad (3.55)$$

We have proved, in Theorem 2.7, that the positive functions $w(x)$ and $-u'(x)$ intersect exactly once on $(0,1)$. By scaling of $w(x)$, we may assume that they intersect at x_0. (Since $w(x)$ is a solution of a linear problem, it is defined up to a constant multiple.) We then have

$$-u' < w \text{ on } (0, x_0), \quad -u' > w \text{ on } (x_0, 1). \qquad (3.56)$$

In view of Lemma 3.3, we conclude

$$\int_0^1 f''(u(x))w^3(x)\, dx < \int_0^1 f''(u(x))w^2(-u_x)\, dx < 0,$$

since, by (3.55) and (3.56), the integrand on the right is pointwise greater than the one on the left. $\diamond$

Remark Recall that in the preceding chapter, we had compared the sign of the integral $\int_0^1 f''(u(x))w^3(x)\, dx$ with that of $\int_0^1 f''(u(x))wu_x^2\, dx$. The present proof is based on the formula (3.53), involving the sign of the integral $\int_0^1 f''(u(x))w^2 u_x\, dx$. In the PDE case, there is no analog of the formula (3.53), and, correspondingly, there are few exact multiplicity results in case $f(0) > 0$.

Proof of the Theorem 3.6 By the Implicit Function Theorem, there is a curve of positive solutions of (3.42), starting at $(\lambda = 0, u = 0)$. This curve continues for increasing λ, until a possible singular solution (λ_0, u_0) is reached, at which point the Crandall-Rabinowitz Theorem 1.2 applies. When $u(0) < \alpha_1$, only turns to the left are possible, since $f(u)$ is convex for $u \in (0, \alpha_1)$. By the Theorem 3.4, when $u(0) = \alpha_1$, the solution is unstable. Hence, a singular solution was reached before $u(0)$ had reached that level, and since only turns to the left are possible when $u(0) < \alpha_1$, it follows that *exactly one* turn has occurred, and at $u(0) = \alpha_1$, the solution curve travels to the left.

We now show that the solution curve keeps traveling to the left, until $u(0)$ increases to the level when only turns to the right are possible. For that we take a close look at the function $h(u) = 2F(u) - uf(u)$. Since

$$h'(u) = f(u) - uf'(u), \quad h''(u) = -uf''(u),$$

it follows that the function $h'(u)$ is decreasing on $(0, \alpha_1)$, and increasing on (α_1, α_2). We have $h'(0) = f(0) > 0$, and so $h'(u)$ can have at most two roots. We claim that it has exactly two roots, denoted u_1 and u_2, with $h'(u)$ being positive on $(0, u_1) \cup (u_2, \bar{u})$, and negative on (u_1, u_2). Indeed, existence of the first root is clear, since $h(0) = 0$ and $h(\alpha_1) < 0$. As for the second root u_2, it exists, because $h(\alpha_2) > 0$ (as follows from our condition (3.35)). So that the function $h(u)$ starts with $h(0) = 0$, it is increasing on $(0, u_1)$, decreasing on (u_1, u_2), with absolute minimum at u_2, and then it increases on the interval $(u_2, \bar{u})$ (see the drawing of $h(u)$). By the Theorem 3.4, the solution curve keeps traveling to the left, while $u(0) \in (\alpha_1, u_2)$. We claim that for $u(0) \in (u_2, \alpha_2)$, the Lemma 3.4 applies. For that we need to check that when $\beta = u_2$, the condition (3.46) holds. Indeed, since $h(u_2) < 0$, we have $f(u_2)u_2 > 2F(u_2)$. Hence

$$I(u_2) = f^2(u_2) - 2F(u_2)f'(u_2) > f^2(u_2) - u_2 f(u_2) f'(u_2) = 0,$$

and the claim follows (observe that $f'(u_2) = \frac{f(u_2)}{u_2} > 0$). By Lemma 3.4, only turns to the right are possible when $u(0) \in (u_2, \alpha_2)$.

Let us now put it all together. We have a curve of solutions, which starts at $(\lambda = 0, u = 0)$. As we travel on this curve, $u(0)$ is always increasing. By the time we reach $u(0) = \alpha_1$ level, the solution curve has made exactly one turn to the left. When $\alpha_1 < u(0) < u_2$, the solution curve travels to the left. When $u(0) > u_2$, only turns to the right are possible, and since when $u(0) \geq \alpha_2$, the solution travels to the right (by the Theorem 3.5), exactly one such turn has occurred. It follows that the solution curve is exactly S-shaped. $\diamondsuit$

Let us consider the problem
$$u'' + \lambda e^{\frac{au}{u+a}} = 0 \quad \text{for } -1 < x < 1, \ u(-1) = u(1) = 0, \qquad (3.57)$$
where a is a constant. This problem is referred to as "the perturbed Gelfand problem", see e.g., J. Bebernes and D. Eberly [17]. In case $a \leq 4$, the problem is easy. In that case, $uf'(u) - f(u) < 0$ for all $u > 0$, and hence all positive solutions are non-singular. This means that the solution curve is monotone, i.e., it continues for all $\lambda > 0$, without any turns. S.-H. Wang [189] has proved existence of a constant a_0, so that for $a > a_0$, the solution curve of (3.57) is exactly S-shaped, i.e., it starts at $\lambda = 0$, $u = 0$, it makes exactly two turns, and then it continues for all $\lambda > 0$, without any more turns. S.-H. Wang [189] gave an approximation of the constant $a_0 \simeq 4.4967$. That paper, as well as all previous ones (notably, K. J. Brown, M. M. A. Ibrahim and R. Shivaji [26]), used time map approach. P. Korman and Y. Li [103] have applied the bifurcation theory approach (i.e., an earlier version

of the Theorem 3.6 above) to this problem, to show the exactness of the S-shaped solution curve, and they also improved the value of the constant to $a_0 \simeq 4.35$, i.e., for $a > a_0$, the solution sets are S-shaped curves. S.-H. Wang [189] has conjectured existence of a critical number $\bar{a}$, so that for $a \leq \bar{a}$ the solution curve is monotone, while for $a > \bar{a}$ the solution curve is exactly S-shaped. P. Korman, Y. Li and T. Ouyang [111] have given a computer assisted proof of the S.-H. Wang's conjecture, by using bifurcation analysis, and these results are presented in the next section. Numerical calculations show that $\bar{a} \simeq 4.07$. Recently, K.-C. Hung and S.-H. Wang [73] have further improved the value of the constant a_0, discussed above, to $a_0 \simeq 4.166$. This value is remarkably close to the numerical approximation of $\bar{a}$.

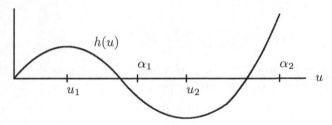

Fig. 3.4　The function $h(u)$

3.5　Computing the location and the direction of bifurcation

Assume that for the problem

$$u''(x) + \lambda f(u(x)) = 0, \quad x \in (-1, 1), \quad u(-1) = u(1) = 0, \tag{3.58}$$

a turn occurs at $u(0) = \alpha$, for which it is necessary that the corresponding linearized problem

$$w''(x) + \lambda f'(u(x))w(x) = 0, \quad x \in (-1, 1), \quad w(-1) = w(1) = 0 \tag{3.59}$$

admits a non-trivial solution (recall that $u(x)$ is then called a singular solution). Observe that $u'(x)$ gives a solution of the equation in (3.59). Since the equation in (3.59) is linear, knowledge of one solution allows one to compute the second one (and hence the general solution). This simple observation provides the basis for the results of this section.

The following result of P. Korman, Y. Li and T. Ouyang [111] provides a way to determine all possible α's, at which bifurcation may occur, i.e., when the corresponding solution of (3.58) is singular. (Recall that the value of $u(0) = \alpha$ uniquely determines the solution pair $(\lambda, u(x))$ of (3.58).) We assume that $u'(1) < 0$ at a singular solution. In case $f(0) \geq 0$, this automatically follows by the Hopf's boundary lemma.

Theorem 3.8. *A positive solution of the problem (3.58), with the maximal value $\alpha = u(0)$, and $u'(1) < 0$, is singular if and only if*

$$G(\alpha) \equiv F(\alpha)^{1/2} \int_0^\alpha \frac{f(\alpha) - f(\tau)}{[F(\alpha) - F(\tau)]^{3/2}} \, d\tau - 2 = 0. \qquad (3.60)$$

Proof: We need to show that the problem (3.59) has a non-trivial solution. Recall that both $u(x)$ and $w(x)$ are even functions. By a direct verification, the function

$$z(x) = -u'(x) \int_x^1 \frac{1}{u'^2(t)} \, dt, \quad x \in (0, 1) \,,$$

satisfies the equation in (3.59). By L'Hospital's rule, we may define $z(0) = \frac{1}{f(u(0))}$. Clearly, $z(1) = 0$. If we also have

$$z'(0) = 0, \qquad (3.61)$$

then the even extension of $z(x)$ to the interval $[-1, 0]$ satisfies $z(-1) = 0$, and it gives us a non-trivial solution of (3.59). Conversely, every non-trivial solution of (3.59) is an even function, and it is a constant multiple of $z(x)$, and hence (3.61) is satisfied. So that the solution $u(x)$ is singular, if and only if (3.61) is satisfied.

Using the equation in (3.58), we compute, for $x \in (0, 1)$,

$$z'(x) = \lambda f(u(x)) \int_x^1 \frac{1}{u'^2(t)} \, dt + \frac{1}{u'(x)} \,. \qquad (3.62)$$

As $x \downarrow 0$, the first term on the right tends to ∞, and the second one to $-\infty$. Therefore, we need to combine them. Write

$$\frac{1}{u'(x)} = -\int_x^1 \frac{d}{dt} \frac{1}{u'(t)} \, dt + \frac{1}{u'(1)} = -\lambda \int_x^1 \frac{f(u(t))}{u'^2(t)} \, dt + \frac{1}{u'(1)} \,,$$

and use this in (3.62), to obtain

$$z'(x) = \lambda \int_x^1 \frac{f(u(x)) - f(u(t))}{u'^3(t)} u'(t) \, dt + \frac{1}{u'(1)} \,. \qquad (3.63)$$

From the energy relation

$$\frac{1}{2} u'^2(t) + \lambda F(u(t)) = \lambda F(\alpha) \,,$$

we express $u'(t) = -\sqrt{2\lambda \left[F(\alpha) - F(u(t)) \right]}$, for $t > 0$, and $u'(1) = -\sqrt{2\lambda F(\alpha)}$. In the integral in (3.63), we make a change of variables $t \to u$, by letting $u = u(t)$, obtaining

$$z'(x) = -\frac{1}{\sqrt{\lambda}} \int_{u(x)}^{0} \frac{f(u(x)) - f(u)}{2^{3/2} \left[F(\alpha) - F(u) \right]^{3/2}} \, du - \frac{1}{\sqrt{2\lambda F(\alpha)}} \, .$$

Taking the limit, as $x \to 0+$ (then $u(x) \to \alpha$, $z'(x) \to 0$ by (3.61)), we obtain (3.60). ◇

Remark. Derivative of the time map gives an alternative way to determine α's, at which bifurcation may occur, see e.g., S.-H. Wang [188]. The function $G(\alpha)$, defined in (3.60), is not proportional to the derivative of the time map, although both functions have the same roots.

Suppose now that $\alpha = u(0)$ satisfies (3.60), so that $(\lambda, u(x))$ is a singular point. Which way does solution curve turn at $(\lambda, u(x))$?

Recall that the sign of the integral

$$I = \int_{0}^{1} f''(u) w^3 \, dx = \frac{1}{2} \int_{-1}^{1} f''(u) w^3 \, dx$$

governs the direction of bifurcation. The following result from P. Korman, Y. Li and T. Ouyang [111], allows one to compute the integral I as a function of $\alpha = u(0)$.

Theorem 3.9. *At any singular solution $u(x)$ of (3.58), with $u(0) = \alpha$, and $u'(1) < 0$,*

$$I = c \int_{0}^{\alpha} f''(u) \left(\int_{u}^{\alpha} f(s) \, ds \right) \left(\int_{0}^{u} \frac{ds}{\left(\int_{s}^{\alpha} f(t) \, dt \right)^{3/2}} \right)^3 \, du, \qquad (3.64)$$

where $c = \frac{1}{(2\lambda)^{7/2}} > 0$.

Proof: At a singular solution, we have just computed the solution of the corresponding linearized problem

$$w(x) = -u'(x) \int_{x}^{1} \frac{1}{u'^2(t)} \, dt \, , \quad \text{for } x > 0 \, .$$

We then have

$$I = -\int_{0}^{1} f''(u(x)) u'^2(x) \left[\int_{x}^{1} \frac{1}{u'^2(t)} \, dt \right]^3 u'(x) \, dx \, . \qquad (3.65)$$

As before, letting $s = u(t)$, and using that $u'(t) = -\sqrt{2\lambda \left[F(\alpha) - F(u(t)) \right]}$,

$$\int_x^1 \frac{1}{u'^2(t)} \, dt = \int_x^1 \frac{1}{u'^3(t)} \, u'(t) \, dt = \int_0^{u(x)} \frac{1}{(2\lambda)^{3/2} \left[F(\alpha) - F(s) \right]^{3/2}} \, ds \, .$$

Using this in (3.65), and making a change of variables $x \to u$, by letting $u = u(x)$, we conclude (3.64). (Clearly, $F(\alpha) - F(s) = \int_s^\alpha f(t) \, dt$.) $\diamondsuit$

The integral in (3.64) is rather involved, but it is quite feasible for numerical computations.

3.5.1   *Sign changing solutions*

Recently, P. Korman and Y. Li [105] had generalized the results of this section to sign-changing solutions, with $n \geq 0$ interior roots, and to solutions of Neumann problem. Here we shall briefly discuss only the sign-changing solutions. The basic idea is the same. If $u(x)$ is a solution of (3.58), then $u'(x)$ gives a solution of the linearized problem (3.59). Let x_0 be any point on the interval $(0, 1)$, such that $u'(x_0) \neq 0$, then one verifies that $u' \int_{x_0}^x \frac{1}{u'^2(t)} \, dt$ gives the second solution of (3.59), and the general solution, on any interval of monotonicity of $u(x)$, is then

$$w = c_1 u' + c_2 u' \int_{x_0}^x \frac{1}{u'^2(t)} \, dt \, .$$

By patching up these general solutions on all intervals of monotonicity of $u(x)$, we can construct non-trivial solutions of the linearized problem (3.59). P. Korman and Y. Li [105] have given a necessary and sufficient condition that a turn occurs at a solution $u(x)$, with n interior roots, and the maximum value equal to β. If a turn does occur, another formula was given in [105], to determine the direction of the turn.

3.6   A class of symmetric nonlinearities

For the autonomous equations both time map analysis, and bifurcation theory apply. If we allow explicit dependence of the nonlinearity on x, i.e., consider

$$u'' + \lambda f(x, u) = 0 \quad \text{for } -1 < x < 1, \quad u(-1) = u(1) = 0, \tag{3.66}$$

then the problem becomes much more complicated. For example, solutions of the corresponding linearized problem need not be of one sign. In several papers of P. Korman and T. Ouyang, which we shall review below, a class of symmetric $f(x, u)$ has been identified, for which the theory of positive solutions is very similar to that for the autonomous case, see e.g., [113], [114] and [115]. Further results in this direction have been given in P. Korman, Y. Li and T. Ouyang [108], and P. Korman and J. Shi [117]. Namely, we assume that $f(x, u) \in C^2 \left([-1, 1] \times \bar{R}_+ \right)$ satisfies

$$f(-x, u) = f(x, u) \quad \text{for all } -1 < x < 1, \text{ and } u > 0, \tag{3.67}$$

$$f_x(x, u) \leq 0 \quad \text{for all } 0 < x < 1, \text{ and } u > 0. \tag{3.68}$$

Under the above conditions the following facts, similar to those for autonomous problems, have been established.

1. Any positive solution of (3.66) is an even function, with $u'(x) < 0$ for all $x \in (0, 1]$. This follows from B. Gidas, W.-M. Ni and L. Nirenberg [63]. (For the one-dimensional problem (3.66) a different proof of the symmetry of solutions is given in P. Korman [78]. It is a little simpler than the moving plane method of [63], and it allows one to relax somewhat the condition (3.68).)

2. Assume, additionally, that $f(x, u) > 0$. Then the maximum value of solution, $\alpha = u(0)$, uniquely identifies the solution pair $(\lambda, u(x))$.

3. Any non-trivial solution of the corresponding linearized problem

$$w'' + \lambda f_u(x, u)w = 0 \quad \text{for } -1 < x < 1, \quad w(-1) = w(1) = 0 \tag{3.69}$$

is of one sign on $(-1, 1)$.

4. The crucial transversality condition of the Crandall-Rabinowitz Theorem 1.2 holds:

$$\int_{-1}^{1} f(x, u)w \, dx \neq 0 \,.$$

5. One can establish stability and instability results, depending only on the maximum value of solution, $u(0) = \alpha$.

As a consequence, many uniqueness and exact multiplicity results were established in the above mentioned papers. We shall review some of them, after we present the basic properties.

Lemma 3.5. *([113]) Under the conditions (3.67) and (3.68), any non-trivial solution of the linearized problem (3.69) is of one sign. Moreover, $w(x)$ is an even function, and it spans the null set of (3.69).*

Proof:　Assume, on the contrary, that $w(x)$ has a root on $[0,1)$, where $u'(x) < 0$ (the case when $w(x)$ vanishes on $(-1,0]$ is similar). We may assume (considering $-w$, if necessary) that there is a subinterval (x_1, x_2), $0 \leq x_1 < x_2 \leq 1$, so that $w(x) > 0$ on (x_1, x_2), and $w(x_1) = w(x_2) = 0$. Using the equations (3.66) and (3.69), we express

$$[u'w' - u''w]' = \lambda f_x w .$$

Integrating this relation over (x_1, x_2),

$$u'(x_2)w'(x_2) - u'(x_1)w'(x_1) = \lambda \int_{x_1}^{x_2} f_x w \, dx.$$

We have a contradiction, since the quantity on the left is positive, while the one on the right is non-positive.

To see that $w(x)$ is even, observe that $w(-x)$ satisfies the same equation as $w(x)$, and hence $w(-x) = aw(x)$, for some constant a (solutions of (3.69) are uniquely defined by the value of $w'(1)$). Setting $x = 0$, we see that $a = 1$.
$\Diamond$

The next lemma is crucial for verifying that the Crandall-Rabinowitz Theorem 1.2 applies at any singular solution.

Lemma 3.6. *Under the conditions (3.67), and (3.68), let $u(x)$ be a singular solution of (3.66), and $w(x) > 0$, a solution of the corresponding linearized problem (3.69). Then we have*

$$\int_0^1 f(x, u)w \, dx > \frac{1}{2\lambda} u'(1)w'(1) > 0. \qquad (3.70)$$

Proof:　By the preceding lemma, we may assume that $w(x) > 0$ on $(-1, 1)$. We then have

$$(u''w - u'w')' = -\lambda f_x w > 0, \quad \text{for } x > 0.$$

So that the function $u''w - u'w'$ is increasing on $(0, 1)$, and then

$$u''w - u'w' < -u'(1)w'(1), \quad \text{for } x > 0.$$

Integrating this over $(0, 1)$, and expressing u'' from the equation (3.66), we conclude (3.70).
$\Diamond$

The following result is taken from P. Korman and J. Shi [117]. Unlike the autonomous case, several conditions are now required.

Theorem 3.10. *(See [117]) In addition to (3.67) and (3.68), assume that*

$$f(x, u) > 0 \quad \text{for all } -1 < x < 1, \text{ and } u > 0. \qquad (3.71)$$

Then the set of positive solutions of (3.66) can be parameterized by their maximum values $u(0)$. (I.e., $u(0)$ uniquely determines the solution pair $(\lambda, u(x))$.)

Proof: Assume, on the contrary, that $v(x)$ is another solution of (3.66), corresponding to some parameter $\mu \geq \lambda$, and $u(0) = v(0)$. The case of $\mu = \lambda$ is not possible, in view of uniqueness of solutions for initial value problems, so we may assume that $\mu > \lambda$. Then $v(x)$ is a supersolution of (3.66), i.e.,

$$v'' + \lambda f(x, v) < 0 \quad \text{for } -1 < x < 1, \quad v(-1) = v(1) = 0. \qquad (3.72)$$

Since $v''(0) < u''(0)$, it follows that $v(x) < u(x)$, for $x > 0$ small. Let $0 < \xi \leq 1$ be the first point where the graphs of $u(x)$ and $v(x)$ intersect (i.e., $v(x) < u(x)$ on $(0, \xi)$). We now multiply the equation (3.66) by u', and integrate over $(0, \xi)$. Since the function $u(x)$ is decreasing on $(0, 1)$, its inverse function exists. Denoting by $x = x_2(u)$, the inverse function of $u(x)$ on $(0, \xi)$, we have

$$\frac{1}{2} u'^2(\xi) + \int_{u(0)}^{u(\xi)} f(x_2(u), u) \, du = 0. \qquad (3.73)$$

Similarly denoting by $x = x_1(u)$ the inverse function of $v(x)$ on $(0, \xi)$, we have from (3.72)

$$\frac{1}{2} v'^2(\xi) + \int_{u(0)}^{u(\xi)} f(x_1(u), u) \, du > 0. \qquad (3.74)$$

Subtracting (3.74) from (3.73), we have

$$\frac{1}{2} \left[u'^2(\xi) - v'^2(\xi) \right] + \int_{u(\xi)}^{u(0)} [f(x_1(u), u) - f(x_2(u), u)] \, du < 0.$$

Noticing that $x_2(u) > x_1(u)$ for all $u \in (u(\xi), u(0))$, and using the condition (3.68), both terms on the left are positive, and we obtain a contradiction.

We consider next the question of stability of positive solutions for a class of symmetric problems

$$u'' + a(x)f(u) = 0, \quad -1 < x < 1, \quad u(-1) = u(1) = 0, \qquad (3.75)$$

with a given even function $a(x)$. We assume that the function $a(x) \in C^1(-1, 1) \cap C[-1, 1]$ satisfies

$$a(x) > 0, \quad a(-x) = a(x), \quad a'(x) < 0 \quad \text{for } x \in (0, 1), \qquad (3.76)$$

while $f(u) \in C^2(\bar{R}_+)$ satisfies

$$f(u) > 0 \quad \text{for } u > 0. \tag{3.77}$$

These assumptions imply that the conditions (3.67), (3.68), and (3.71) hold, and hence the preceding results of this section apply. We denote by $(\mu_1, w(x) > 0)$ the principal eigenpair of the eigenvalue problem for the corresponding linearized equation

$$w'' + a(x)f'(u)w + \mu w = 0, \quad -1 < x < 1, \quad w(-1) = w(1) = 0. \tag{3.78}$$

We see, similarly to Lemma 3.5, that $w(x)$ is an even function. Again, we define $h(u) = 2F(u) - uf(u)$, and also $p(x) = u'(x)w'(x) - u''(x)w(x) = u'(x)w'(x) + a(x)f(u(x))w(x)$. Observe that $p(x)$ is an even function. We shall need the following simple lemma from P. Korman [94].

Lemma 3.7. *In addition to the conditions (3.76) and (3.77), assume that either*

$$\mu_1 \leq 0, \tag{3.79}$$

or

$$\mu_1 > 0, \quad \text{and } f'(u) \geq 0 \text{ for } u > 0. \tag{3.80}$$

Then

$$p(x) > 0, \quad \text{for all } x \in (-1, 1). \tag{3.81}$$

Proof: We have $p(1) = u'(1)w'(1) > 0$, and in case $\mu_1 \leq 0$,

$$p'(x) = -\mu_1 u'(x)w(x) + a'(x)f(u(x))w(x) < 0, \quad \text{for } x > 0,$$

and (3.81) follows. In case the condition (3.80) holds, we see from (3.78) that the even function $w(x) > 0$ is concave, which implies that $w'(x) < 0$ for all $x > 0$, and hence positivity of the even function $p(x)$ follows directly from its definition. $\diamond$

Similarly to the autonomous case, we have the following result from P. Korman [94]. The instability part was proved previously by P. Korman and J. Shi [117], using a different method (see an exposition in P. Korman [92]).

Theorem 3.11. *Assume that the conditions (3.76) and (3.77) hold.*
(i) *If the condition*

$$h(\alpha) > h(u), \quad \text{for all } 0 < u < \alpha \tag{3.82}$$

holds, then the positive solution of (3.75), with $u(0) = \alpha$, is stable.
(ii) *If the condition*

$$h(\alpha) < h(u), \quad \text{for all } 0 < u < \alpha \tag{3.83}$$

holds, and $f'(u) \geq 0$ for $u \in (0, \alpha)$, then the positive solution of (3.75), with $u(0) = \alpha$, is unstable.

Proof: Using the equations (3.75) and (3.78), we have

$$I \equiv \int_{-1}^{1} a(x)\left(f(u) - uf'(u)\right) w \, dx = \mu_1 \int_{-1}^{1} uw \, dx. \quad (3.84)$$

We have,

$$I = \int_{-1}^{1} \frac{d}{dx} \left[h(u) - h(\alpha)\right] \frac{a(x)w(x)}{u'(x)} \, dx \quad (3.85)$$

$$= -\int_{-1}^{1} \left[h(u) - h(\alpha)\right] \frac{a'(x)u'(x)w(x) + a(x)p(x)}{u'^2} \, dx.$$

In case **(i)** holds, let us assume, on the contrary, that the solution is not stable, i.e., $\mu_1 \leq 0$. Then $I \leq 0$ from (3.84), while $I > 0$ from (3.85), since $p(x) > 0$ by the above lemma, a contradiction.

If case **(ii)** holds, assume, on the contrary, that the solution fails to be unstable, i.e., $\mu_1 \geq 0$. Then $I \geq 0$ from (3.84), while $I < 0$ from (3.85), again resulting in a contradiction. $\diamond$

Example The solution curve, in $(\lambda, u(0))$ plane, for the problem

$$u'' + \lambda a(x)(2u + \sin u) = 0, \quad -1 < x < 1, \quad u(-1) = u(1) = 0,$$

has infinitely many turns, assuming that the function $a(x)$ satisfies the conditions (3.76). Indeed, the function $f(u) = 2u + \sin u$ is positive and increasing for $u > 0$, while for $h(u) = 2 - 2\cos u - u \sin u$, there is a sequence $\{\alpha_n\} \to \infty$, so that $h(u) > h(\alpha_n)$ for all $u \in (0, \alpha_n)$, and another sequence $\{\beta_n\} \to \infty$, so that $h(u) < h(\beta_n)$ for all $u \in (0, \beta_n)$. Solutions with $u(0) = \alpha_n$ are unstable, and the ones with $u(0) = \beta_n$ are stable.

For convex $f(u)$, we have the following exact multiplicity result from P. Korman and J. Shi [117]. It extends the corresponding result in P. Korman and T. Ouyang [113], by not restricting the behavior of $f(u)$ at infinity.

Theorem 3.12. *For the problem (3.75), assume that $a(x)$ satisfies the conditions (3.76). Assume that $f(u) \in C^2[0, \infty)$, $f(u) > 0$, $f'(u) > 0$, and $f''(u) > 0$ for all $u > 0$, while $h(\alpha) \leq 0$ for some $\alpha > 0$. Then there exist two constants $0 \leq \bar{\lambda} < \lambda_0$, so that the problem (3.75) has no solution for $\lambda > \lambda_0$, exactly one solution at $\lambda = \lambda_0$, exactly two solutions for $\bar{\lambda} < \lambda < \lambda_0$, and in case $\bar{\lambda} > 0$ it has exactly one solution for $0 < \lambda < \bar{\lambda}$. Moreover, all solutions lie on a unique smooth solution curve. If, in addition, we assume that $\lim_{u \to \infty} \frac{f(u)}{u} = \infty$, then $\bar{\lambda} = 0$.*

We discuss next the cubic problems

$$u'' + \lambda(u - a(x))(u - b(x))(c(x) - u) = 0 \quad \text{for } -1 < x < 1, \quad (3.86)$$

$$u(-1) = u(1) = 0,$$

with given even functions $0 \leq a(x) \leq b(x) \leq c(x)$. As we mentioned above, for constant a, b and c, the exact multiplicity question has been fully settled only recently by P. Korman, Y. Li and T. Ouyang [111], via a computer assisted proof. One can expect that under some conditions the same global picture holds for variable coefficients. This was established for several special cases. P. Korman, Y. Li and T. Ouyang [108] have given an exact multiplicity result in case $a = b = 0$. P. Korman and T. Ouyang [115] had done the same in case $a = 0$, and P. Korman and T. Ouyang [112] had given an exact multiplicity result in case when $a > 0$ is a constant. We review these results next.

Theorem 3.13. *([108]) Consider the problem*

$$u'' + \lambda u^2(c(x) - u) = 0 \quad for -1 < x < 1, \ u(-1) = u(1) = 0. \quad (3.87)$$

Assume that the given function $c(x) \in C^3(-1,1) \cap C[-1,1]$ *satisfies*

$$c(x) > 0, \ c(-x) = c(x), \ c'(x) < 0 \quad for \ x \in (0,1),$$

$$c''(x) < 0, \ c'''(x) \leq 0 \quad for \ x \in (0,1),$$

$$c(1) \geq \frac{2}{3}c(0).$$

Then there is a critical $\lambda_0 > 0$, *so that the problem (3.87) has no non-trivial solutions for* $\lambda < \lambda_0$, *it has exactly one non-trivial solution for* $\lambda = \lambda_0$, *and exactly two non-trivial solutions for* $\lambda > \lambda_0$. *Moreover, all solutions lie on a single smooth solution curve, which for* $\lambda > \lambda_0$ *has two branches,* $u^-(x,\lambda) < u^+(x,\lambda)$, *with* $u^+(x,\lambda)$ *strictly monotone increasing in* λ, $u^-(x,\lambda)$ *strictly monotone increasing in* λ, *and* $\lim_{\lambda \to \infty} u^+(x,\lambda) = c(x)$, $\lim_{\lambda \to \infty} u^-(x,\lambda) = 0$, *for all* $x \in (-1,1)$.

For example, $c(x) = A - x^2$, with a constant $A \geq 3$, satisfies all of the above conditions. Observe that by the maximum principle, non-trivial solutions of (3.87) are in fact positive, satisfying $0 < u(x) < c(x)$ for all $x \in (-1,1)$.

We consider, next, the problem

$$u'' + \lambda u(u - a(x))(b(x) - u) = 0 \quad for -1 < x < 1, \quad (3.88)$$

$$u(-1) = u(1) = 0.$$

Here $f(x,u) = u(u - a(x))(b(x) - u) = -u^3 + \alpha(x)u^2 - \beta(x)u$, where we denote $\alpha(x) = a(x) + b(x)$ and $\beta(x) = a(x)b(x)$. We assume that the functions $a(x)$ and $b(x)$ are even, with

$$0 < a(x) < b(x), \quad for \ x \in (0,1). \quad (3.89)$$

The functions $\alpha(x)$ and $\beta(x)$ are also even, and we assume that they satisfy

$$\alpha'(x) < 0, \quad \text{and} \quad \beta'(x) > 0, \quad \text{for } x \in (0,1). \tag{3.90}$$

Then $f_x(x, u) < 0$, for $x \in (0, 1)$ and $u > 0$, and so by B. Gidas, W.-M. Ni and L. Nirenberg [63], any positive solution of (3.88) is an even function, with $xu'(x) < 0$ for all $x \neq 0$. As before, any non-trivial solution of (3.88) is in fact positive.

Theorem 3.14. *([115]) In addition to the conditions (3.89) and (3.90), assume the following*

$$\alpha'''(x) < 0, \quad \text{for } x \in (0,1),$$

$$\alpha(x) - \sqrt{\alpha^2(x) - 3\beta(x)} < \alpha(1), \quad \text{for } x \in (0,1),$$

$$\frac{1}{2}\alpha(0) < \frac{\alpha(x) + \sqrt{\alpha^2(x) - 3\beta(x)}}{3}, \quad \text{for } x \in (0,1).$$

Let us denote $F(x,u) = \int_0^u f(x,z)\,dz$, *and assume finally that*

$$\int_0^1 F(x, b(x))\,dx > 0.$$

Then there is a critical $\lambda_0 > 0$, *so that the problem (3.88) has no positive solutions for* $\lambda < \lambda_0$, *it has exactly one positive solution for* $\lambda = \lambda_0$, *and exactly two positive solutions for* $\lambda > \lambda_0$. *Moreover, all solutions lie on a single smooth parabola-like solution curve, with a single turn to the right.*

Example Let $a(x) = A + Bx^2$, $b(x) = C - Dx^2$. One verifies that all of the above conditions are satisfied, provided that $D > B$, and C is large enough.

P. Korman and T. Ouyang [113] have considered a class of *indefinite* problems

$$u''(x) + \lambda u(x) + h(x)u^p(x) = 0, \quad -1 < x < 1, \quad u(-1) = u(1) = 0. \tag{3.91}$$

Here $p > 1$, and λ a real parameter. The given function $h(x)$ is assumed to be even, and it changes sign on $(-1, 1)$. By using bifurcation analysis, as well as the earlier work of T. Ouyang [142], [143], it was possible to give an exhaustive description of the set of positive solutions of (3.91). We describe this result, after some definitions.

We denote by $\phi_1 = \cos \frac{\pi}{2}x > 0$, the principal Dirichlet eigenfunction of $-u''$ on $(-1, 1)$, corresponding to the principal eigenvalue $\lambda_1 = \frac{\pi^2}{4}$. We assume that $h(x) \in C^1(-1, 1) \cap C^0[-1, 1]$ is an even function, and moreover

$$h(0) > 0, \quad \text{and} \quad h'(x) < 0 \text{ for } x \in (0,1), \tag{3.92}$$

$$\int_{-1}^{1} h(x)\phi_1^{p+1}(x)\,dx < 0. \tag{3.93}$$

(The last assumption implies that $h(x)$ changes sign.)

Theorem 3.15. *([113]) Assume that the conditions (3.92) and (3.93) hold for the problem (3.91). Then there is a critical λ_0, $\lambda_0 > \lambda_1$, so that for $-\infty < \lambda \le \lambda_1$ the problem (3.91) has exactly one positive solution, it has exactly two positive solutions for $\lambda_1 < \lambda < \lambda_0$, exactly one at $\lambda = \lambda_0$, and no positive solutions for $\lambda > \lambda_0$. Moreover, all positive solutions lie on a unique continuous in λ curve, which bifurcates from zero at $\lambda = \lambda_1$ (to the right), and continues without any turns to $\lambda = \lambda_0$, at which point it turns to the left, and then continues without any more turns for all $-\infty < \lambda \le \lambda_0$. We also have: $u(0) = \max u(x) \to \infty$, as $\lambda \to -\infty$. (See the bifurcation diagram.)*

This appears to be the only known exact multiplicity result for indefinite problems.

Finally, we indicate an extension. Consider a problem with a variable diffusion coefficient

$$(a(x)u')' + \lambda f(u) = 0 \quad \text{for } -1 < x < 1, \ u(-1) = u(1) = 0. \tag{3.94}$$

We assume that a given function $a(x) \in C^1[-1, 1]$ is even, and it satisfies

$$a(x) > 0 \quad \text{and} \quad xa'(x) \le 0, \quad \text{for } x \in [-1, 1]. \tag{3.95}$$

We perform a change of variables $x \to s$, given by

$$s = \int_0^x \frac{dt}{a(t)}.$$

If we denote by $s_0 = \int_0^1 \frac{dt}{a(t)}$, then this transformation gives a one-to-one map of the interval $(-1, 1)$ onto $(-s_0, s_0)$. Moreover, $s > 0 \ (< 0)$ if and only if $x > 0 \ (< 0)$. The problem (3.94) transforms into

$$u_{ss} + \lambda a(x(s))f(u) = 0 \quad \text{for } -s_0 < x < s_0, \ u(-s_0) = u(s_0) = 0. \tag{3.96}$$

Observe that the function $s = s(x)$ is odd, and hence its inverse $x = x(s)$ is also odd, and then $a(x(s))$ is even. In view of (3.95),

$$\frac{d}{ds}a(x(s)) = a'(x(s))\,a(x(s)) \le 0 \ (\ge 0) \quad \text{if } s > 0 \ (< 0).$$

If we now assume that $f(u) > 0$ for $u > 0$, then the problem (3.96) satisfies the conditions (3.67) and (3.68). Hence, we can translate the results of this section to the problem (3.94).

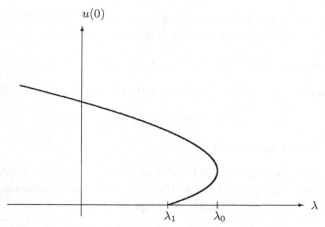

Fig. 3.5   Bifurcation diagram for the indefinite problem (3.91)

3.7   General nonlinearities

Without the symmetry assumptions on $f(x, u)$, the problem is much harder. We restrict to a subclass of such problems, i.e., we now consider positive solutions of the boundary value problem

$$u'' + \lambda \alpha(x)f(u) = 0 \quad \text{for } a < x < b, \quad u(a) = u(b) = 0, \tag{3.97}$$

on an arbitrary interval (a, b). We assume that $f(u)$ and $\alpha(x)$ are positive functions of class C^2, i.e., $f(u) \in C^2(\bar{R}_+)$, $\alpha(x) \in C^2([a, b])$, and

$$f(u) > 0 \text{ for } u > 0, \quad \alpha(x) > 0 \text{ for } x \in [a, b]. \tag{3.98}$$

As before, it will be crucial for bifurcation analysis to prove positivity property for the corresponding linearized problem

$$w'' + \lambda \alpha(x)f'(u)w = 0 \quad \text{for } a < x < b, \quad w(a) = w(b) = 0. \tag{3.99}$$

The following result was proved in P. Korman and T. Ouyang [112], although here we follow a little different exposition from P. Korman [92].

Lemma 3.8. *In addition to the conditions (3.98), assume that*

$$\frac{3}{2}\frac{\alpha'^2}{\alpha} - \alpha'' < 0 \quad \text{for all } x \in (a, b). \tag{3.100}$$

If the linearized problem (3.99) admits a non-trivial solution, then we may assume that $w(x) > 0$ on (a, b).

Proof: We shall use a "test function" $z(x) \equiv g(x)u'(x)$, with $g(x)$ to be chosen shortly. One verifies that $z(x)$ satisfies the equation

$$z'' + \lambda\alpha(x)f'(u)z = g''(x)u'(x) - \lambda\left(2g'(x)\alpha(x) + \alpha'(x)g(x)\right)f.$$

We now choose $g(x) = \alpha(x)^{-1/2}$. Then $2g'(x)\alpha(x) + \alpha'(x)g(x) = 0$, while

$$g''(x) < 0 \ \text{for} \ a < x < b, \tag{3.101}$$

in view of the assumption (3.100).

Notice that any positive solution of (3.97) is a concave function, and hence it has only one critical point, the point of global maximum. Let x_0 be the point of maximum of $u(x)$. We have

$$z'' + \lambda\alpha(x)f'(u)z = g''(x)u'(x), \tag{3.102}$$

with the right hand side negative on (a, x_0), and positive on (x_0, b). This will make it impossible for $w(x)$ to vanish inside (a, b). Indeed, if we assume that $w(x)$ vanishes on, say (x_0, b), we could find two consecutive roots of $w(x)$, x_1 and x_2, with $x_0 \leq x_1 < x_2 \leq b$, so that $w(x_1) = w(x_2) = 0$, while $w(x) > 0$ on (x_1, x_2). We now multiply the equation (3.102) by $w(x)$, the equation (3.99) by $z(x)$, subtract and integrate over (x_1, x_2), obtaining

$$-g(x_2)u'(x_2)w'(x_2) + g(x_1)u'(x_1)w'(x_1) = \int_{x_1}^{x_2} g''(x)u'(x)w(x) \, dx.$$

We have a contradiction, since the quantity on the left is negative, and the integral on the right is positive. $\diamond$

Remarks

(1) Recall the Schwarzian derivative, prominent in Complex Analysis and Dynamical Systems

$$(Sf)(x) = \frac{f'''(x)}{f'(x)} - \frac{3}{2}\left(\frac{f''(x)}{f'(x)}\right)^2.$$

If one denotes $A(x) = \int \alpha(x) \, dx$, then our condition (3.100) says that the Schwarzian derivative of $A(x)$ is positive.

(2) Recall that semilinear equations on an annulus in R^n, $n > 2$, can be reduced by a standard change of variables to the problem (3.97), with $\alpha(x) = x^{-2k}$, and $k = 1 + \frac{1}{n-2}$. One sees that our condition (3.100) narrowly misses this kind of functions.

We shall present another result on positivity of $w(x)$ from P. Korman [92], after we prove a simple lemma. Under our conditions, any solution $u(x)$ of (3.97) is concave, and hence its only critical point is the unique point of maximum.

Lemma 3.9. *Assuming the condition (3.98) holds, let x_0 be the unique point of maximum of a positive solution $u(x)$ of (3.97). Assume that*

$$\alpha'(x) < 0 \quad on \ (a, b). \tag{3.103}$$

Then we have the following two conclusions.

(i) *If the corresponding linearized problem (3.99) admits a non-trivial solution $w(x)$, then this solution $w(x)$ cannot vanish inside (x_0, b).*

(ii) $x_0 \in (a, \frac{a+b}{2})$, *i.e., the solution $u(x)$ is "skewed to the left".*

Proof: (i) Assuming the contrary, let γ be the largest root of $w(x)$ on (x_0, b), and assume that $w(x) > 0$ on (γ, b). (The number of roots of $w(x)$ inside (a, b) is at most finite, as follows by Sturm's comparison theorem, since both functions $f'(u(x))$ and $\alpha(x)$ are bounded on $[a, b]$, and λ is fixed. Hence, there is a largest root γ.) Differentiate the equation (3.97)

$$u_x'' + \lambda\alpha(x)f'(u)u_x + \lambda\alpha'(x)f(u) = 0. \tag{3.104}$$

Multiplying the equation (3.104) by $w(x)$, the equation (3.99) by $u'(x)$, subtracting and integrating, we have

$$-u'(b)w'(b) + u'(\gamma)w'(\gamma) + \lambda \int_\gamma^b \alpha'(x)f(u(x))w(x)\, dx = 0.$$

This results in a contradiction, since all terms on the left are negative.

(ii) If we have $x_1 < x_0 < x_2$, and $u(x_1) = u(x_2)$, then we claim

$$u'(x_1) > |u'(x_2)|. \tag{3.105}$$

Assuming the claim for the moment, we conclude that the function $u(x)$ climbs faster toward its maximum at x_0, than it descends afterwards. Hence $u(x)$ is "tilted" to the left, and so its maximum x_0 occurs in the first half of the interval (a, b).

To prove the claim, we multiply the equation (3.97) by u', and integrate over (x_1, x_0), obtaining

$$-\frac{1}{2}u'^2(x_1) + \lambda \int_{u_1}^{u_0} \alpha(x_1(u))f(u)\, du = 0,$$

where we denote $u_1 = u(x_1)$, $u_0 = u(x_0)$, and $x_1(u)$ the inverse function of $u(x)$ on the interval (x_1, x_0). Similarly, multiplying the equation (3.97) by u', and integrating over (x_0, x_2),

$$\frac{1}{2}{u'}^2(x_2) + \lambda \int_{u_0}^{u_1} \alpha(x_2(u))f(u)\, du = 0,$$

where $x_2(u)$ denotes the inverse function of $u(x)$ on the interval (x_0, x_2). Adding,

$$\frac{1}{2}{u'}^2(x_2) - \frac{1}{2}{u'}^2(x_1) + \lambda \int_{u_1}^{u_0} \left[\alpha(x_1(u)) - \alpha(x_2(u))\right] f(u)\, du = 0 \,.$$

Since $x_2(u) > x_1(u)$ for all $u \in (u_1, u_0)$, the assumption (3.103) implies that the integral term is positive, and hence the claim (3.105) follows. $\diamond$

Remarks

(1) Combining both statements of the lemma, we conclude that any solution $w(x)$ of the linearized problem (3.99) cannot vanish on $(\frac{a+b}{2}, b)$.
(2) For the first conclusion of this lemma, it suffices to assume that $\alpha'(x) < 0$ on (x_0, b). If $\alpha'(x) > 0$ on (a, x_0), then a similar proof shows that $w(x)$ cannot vanish inside (a, x_0).

Lemma 3.10. *Assume that the condition (3.98) holds, and, in addition, assume that*

$$\alpha'(x) < 0 \quad on \ (a,b), \tag{3.106}$$

and

$$2\alpha(x) + (x-a)\alpha'(x) > 0 \quad on \ (a, \tfrac{a+b}{2}). \tag{3.107}$$

If the linearized problem (3.99) admits a non-trivial solution, then we may assume that $w(x) > 0$ on (a,b).

Proof: Let x_0 be the unique point of maximum of the solution $u(x)$. By Lemma 3.9, $w(x)$ cannot vanish on (x_0, b). Assuming that $w(x)$ vanishes on $(a, x_0]$, let $\gamma \in (a, x_0]$ be the first root of $w(x)$, and we may assume that $w(x) > 0$ on (a, γ). We consider the function $\zeta(x) = (x - a)\left[u'(x)w'(x) + \lambda\alpha(x)f(u(x))w(x)\right] - u'(x)w(x)$, a variation on the function, introduced by M. Tang [182], which we had used before. One computes

$$\zeta'(x) = \lambda \left[2\alpha(x) + (x-a)\alpha'(x)\right] f(u)w. \tag{3.108}$$

Integrating this relation over (a, γ),

$$(\gamma - a)u'(\gamma)w'(\gamma) = \lambda \int_a^\gamma [2\alpha(x) + (x - a)\alpha'(x)]\, f(u)w \, dx.$$

We have a contradiction, since the quantity on the left is negative, while the integral on the right is positive. $\diamond$

Example We now consider the case of a thin annulus, justifying the Theorem 2.52 from the preceding chapter. We consider the case $n > 2$, and the case $n = 2$ is similar. Recall that by a change of variables, we had reduced the annulus to the problem (3.97) on an interval (a, b), with $0 < a < b$, $u = u(s)$, $\alpha(s) = (n-2)^{-2}s^{-2k}$, and $k = 1 + \frac{1}{n-2}$. Clearly $\alpha'(s) < 0$, while the condition (3.107) simplifies to

$$s < (n-1)a \quad \text{for } s \in (a, \tfrac{a+b}{2}).$$

For that it suffices to have $\frac{a+b}{2} < (n-1)a$, i.e.,

$$b < (2n - 3)a.$$

Recall that we made a change of variables $s = r^{2-n}$, and we had $a = B^{2-n}$, and $b = A^{2-n}$. Then for the original annulus, we need

$$B \leq c_n A, \quad \text{with } c_n = (2n - 3)^{\frac{1}{n-2}}.$$

That was our condition for a "thin annulus".

We now have all of the ingredients in place to prove uniqueness and exact multiplicity results. For example, we can prove the following theorem.

Theorem 3.16. *For the problem (3.97) assume that the condition (3.98) holds, and that either the condition (3.100) holds, or the conditions (3.106) and (3.107) hold. In addition, assume that $f''(u) > 0$ for all $u > 0$, and $\lim_{u \to \infty} \frac{f(u)}{u} = \infty$. Then there is a critical $\lambda_0 > 0$, so that the problem (3.97) has exactly two positive solutions for $0 < \lambda < \lambda_0$, exactly one positive solution at $\lambda = \lambda_0$, and no positive solutions for $\lambda > \lambda_0$. Moreover, all solutions lie on a unique smooth solution curve, which starts at $(\lambda = 0, u = 0)$, bends back at $\lambda = \lambda_0$, and tends to infinity as $\lambda \downarrow 0$.*

Proof: The proof follows the usual pattern, except for proving the uniqueness of the solution curve (since here the maximum value of a solution no longer identifies the pair $(\lambda, u(x))$). However, if another solution curve existed, one of its ends would have to go through the origin $(\lambda = 0, u = 0)$, contradicting the uniqueness of solutions near regular points, which follows by the Implicit Function Theorem. (Since $w > 0$ at the turning point, one of the branches is increasing in λ, i.e., it is decreasing, as $\lambda \downarrow 0$. By the maximum principle, the monotonicity is preserved along the branch, and hence this branch must go into the origin, as $\lambda \to 0$.) $\diamond$

Example The theorem applies to the problem

$$u'' + \lambda \alpha(x)e^u = 0, \quad a < x < b, \quad u(a) = u(b) = 0,$$

if $\alpha(x) > 0$ satisfies either the condition (3.100), or the conditions (3.106) and (3.107), giving us an exact multiplicity result.

For the autonomous problems we made use of the following two general facts: the maximum value of solution uniquely identifies the pair $(\lambda, u(x))$, and any non-trivial solution for the linearized problem is of one sign. We have seen that both of these properties extend, under some conditions, to symmetric problems. The following two examples show that both of these properties may fail for general equations.

Example The problem

$$u'' + \lambda a(x)u^p = 0 \quad \text{for } 0 < x < 1, \quad u(0) = u(1) = 0,$$

with given sign-changing function $a(x)$ may have multiple positive solutions, see M. Gaudenzi, P. Habets, and F. Zanolin [62]. Let $u_0(x)$ and $v_0(x)$ be two different positive solutions at $\lambda = 1$. Then, $\frac{1}{\lambda^{\frac{1}{p-1}}}u_0(x)$ and $\frac{1}{\lambda^{\frac{1}{p-1}}}v_0(x)$ give two different families of positive solutions, with maximum values changing from zero to infinity. Hence, to each maximum value of solution correspond at least two solution pairs $(\lambda, u(x))$.

Example The Theorem 1.10 in W.-M. Ni and R. Nussbaum [138] shows that there are at least three positive radial solutions for the problem

$$\Delta u + u^p + \epsilon u^q = 0 \quad \text{for } x \in \Omega, \quad u = 0 \text{ on } \partial\Omega, \tag{3.109}$$

where Ω is an annulus, $\Omega = \{x \in R^n \,|\, a < |x| < b\}$, provided that $1 < p < \frac{n+2}{n-2} < q$, and both $\epsilon > 0$, and $a > 0$ are sufficiently small. Recall that the problem (3.109), for radial solutions, may be transformed by a change of variables $s = r^{n-2}$, $n \geq 3$, to

$$u'' + \alpha(s)\left(u^p + \epsilon u^q\right) = 0 \quad \text{for } A < s < B, \quad u(A) = u(B) = 0, \tag{3.110}$$

where $\alpha(s) = (n-2)^{-2}s^{-2k}$, $A = b^{2-n}$, and $B = a^{2-n}$, so that the problem (3.110) has at least three positive solutions. We now deform the problem (3.110) to the autonomous one, by considering

$$u'' + \left(\mu\alpha(s) + 1 - \mu\right)\left(u^p + \epsilon u^q\right) = 0 \quad \text{for } A < s < B, \tag{3.111}$$
$$u(A) = u(B) = 0,$$

where $0 \leq \mu \leq 1$. When $\mu = 1$, we have the problem (3.110), while at $\mu = 0$, we have an autonomous problem with exactly one positive solution (observe that here $f'(u) > \frac{f(u)}{u}$, for $u > 0$). We now continue the three positive solutions of (3.111), as μ decreases from 1 to 0. We will have to encounter singular solutions during this continuation process (otherwise, we would get 3 positive solutions of the autonomous problem, a contradiction). At a singular solution, the non-trivial solution of the corresponding linearized problem would have to be sign-changing, by Sturm's comparison theorem (again, using that $f'(u) > \frac{f(u)}{u}$, for $u > 0$).

3.8 Infinitely many curves with pitchfork bifurcation

We study positive, negative and sign-changing solutions of a family of two-point boundary value problems

$$u'' + \lambda f(u) = 0 \text{ for } x \in (0, 1), \quad u(0) = u(1) = 0, \qquad (3.112)$$

depending on a positive parameter λ. We assume that the nonlinearity $f(u) \in C^2(R)$ satisfies the following conditions:

$$f(0) < 0, \qquad (3.113)$$

$$\lim_{u \to \pm\infty} \frac{f(u)}{u} = \infty. \qquad (3.114)$$

We also assume that $f(u)$ changes sign exactly once, i.e., there exists an $u_0 > 0$, such that

$$f(u) > 0 \text{ for } u > u_0, \quad f(u) < 0 \text{ for } -\infty < u < u_0. \qquad (3.115)$$

For example, $f(u) = u^3 - a$, with a constant $a > 0$, satisfies all of these conditions. We shall show that the problem has infinitely many solution curves with a pitchfork bifurcation, and all of these curves tend to infinity as $\lambda \to 0$, so that for small $\lambda > 0$, infinitely many solutions are present. The results of this section are based on P. Korman [84].

Notice that any solution of (3.112) is symmetric between any two of its roots, i.e., if $u(a) = u(b) = 0$ and $u(x)$ is, say, positive over the subinterval $(a, b) \in (0, 1)$ then $u'(\frac{a+b}{2}) = 0$, and $u(x) = u(a + b - x)$, for all $x \in (a, b)$, with $u'(x) < 0$ for all $x \in (\frac{a+b}{2}, b)$. This is because any solution is symmetric with respect to any of its stationary points, which follows by uniqueness of solution for initial value problems. It also follows that no solution can have local minimums (maximums) in the region where it is positive (negative).

We continue to use alternatively the notation $u(x, \lambda) \equiv u(x)$, and $u_x(x, \lambda) = u'(x, \lambda)$ to denote solution branches and derivatives. We begin with the following general observation.

Lemma 3.11. *The nodal structure of any solution branch of (3.112) is preserved for all λ, unless $u_x(0, \lambda) = 0$ for some λ. More precisely, if $u_x(0, \lambda) \neq 0$ for all $\lambda \in [\lambda_1, \lambda_2]$, then $u(x, \lambda_1)$ and $u(x, \lambda_2)$ have the same number of roots, and the same order of regions where solution is positive or negative.*

Proof: By the above remarks, the only way a new zero point can be created, is for solution curve to develop a zero slope at an existing root, and then a new root appearing nearby, when the parameter λ is varied. However, by symmetry (or using energy), having zero slope at any root would imply zero slope at all roots, i.e., $u_x(0, \lambda) = 0$, contrary to our assumption. $\diamond$

The following lemma is true for any solution $u(x)$, positive or sign-changing.

Lemma 3.12. *Let $u(x, \lambda_0) \in C^3(0, 1) \cap C[0, 1]$ be a singular solution of (3.112), i.e., the corresponding linearized problem*

$$w'' + \lambda_0 f'(u)w = 0 \text{ for } x \in (0, 1), \quad w(0) = w(1) = 0 \qquad (3.116)$$

has a nontrivial solution. Assume that $u'(1, \lambda_0) \neq 0$ (or $u'(0, \lambda_0) \neq 0$). Then

$$\int_0^1 f(u)w \, dx \neq 0. \qquad (3.117)$$

Proof: From the equations (3.112) and (3.116),

$$u'w' - u''w = \text{constant} = u'(1)w'(1). \qquad (3.118)$$

Since we may assume that $w'(1) \neq 0$ (otherwise $w(x)$ is trivial), we see that the constant in (3.118) is non-zero. Integrating (3.118), we conclude (3.117). $\diamond$

The following lemma addresses the complementary situation, when $u'(1, \lambda_0) = 0$. It gives explicitly the solution of the equation for $u_\lambda(x, \lambda)$,

$$u_\lambda'' + \lambda f'(u)u_\lambda + f(u) = 0 \text{ for } x \in (0, 1), \quad u_\lambda(0) = u_\lambda(1) = 0, \quad (3.119)$$

at $\lambda = \lambda_0$, i.e., the "tangential direction" on the solution curve. Observe that the solution satisfying $u'(1, \lambda_0) = 0$, is a singular solution, since $u'(x, \lambda_0)$ gives us a non-trivial solution of the linearized problem (3.116).

Lemma 3.13. *Assume that $u(x, \lambda) \in C^3(0, 1) \cap C[0, 1]$ is a curve of solutions of (3.112), which are symmetric with respect to $x = \frac{1}{2}$ near $\lambda = \lambda_0$, and such that $u_x(1, \lambda_0) = 0$. Then the problem (3.119) is uniquely solvable at $\lambda = \lambda_0$, and in fact*

$$u_\lambda(x, \lambda_0) = \frac{1}{2\lambda_0}(x - \frac{1}{2})u_x(x, \lambda_0), \quad \text{for all } x \in (0, 1). \tag{3.120}$$

Proof: One easily checks that the function $w(x) \equiv 2\lambda_0 u_\lambda(x, \lambda_0) - (x - \frac{1}{2}) u_x(x, \lambda_0)$ satisfies

$$w'' + \lambda_0 f'(u)w = 0 \text{ on } (0, 1), \quad w(0) = w(1) = 0. \tag{3.121}$$

This function is even with respect to $x = \frac{1}{2}$, while the one-dimensional null-space of the problem (3.121) is spanned by the odd, with respect to $x = \frac{1}{2}$, function $u_x(x, \lambda_0)$. It follows that $w \equiv 0$, and the proof follows. $\diamondsuit$

With $F(u) = \int_0^u f(u) \, du$, we define the point $\theta > u_0$, by the relation $F(\theta) = \int_0^\theta f(u) \, du = 0$. It turns out that pitchfork bifurcation occurs at each of the following sequence of parameter values

$$\lambda_k = 2k^2 \left(\int_0^\theta \frac{du}{\sqrt{-F(u)}} \right)^2, \quad \text{for } k = 1, 2 \ldots. \tag{3.122}$$

At $\lambda = \lambda_k$ the problem (3.112) has a non-negative *touchdown* solution, with $k - 1$ roots of zero slope, as described in the following lemma.

Lemma 3.14. *Let λ_k be defined by (3.122). At $\lambda = \lambda_1$ the problem (3.112) has a positive solution $u_1(x)$, with $u_1'(0) = 0$ and $u_1(\frac{1}{2}) = \theta$. At $\lambda = \lambda_k$, $k \geq 2$, the problem (3.112) has a non-negative solution $u_k(x)$ with $k - 1$ interior zeros, $\frac{1}{k}, \ldots, \frac{k-1}{k}$, and $u_k'(0) = u_k'(\frac{i}{k}) = 0$, $i = 1, \ldots, k$. The first point of maximum of $u_k(x)$ is at $x = \frac{1}{2k}$, with $u_k(\frac{1}{2k}) = \theta$.*

Proof: We begin by establishing the existence of solutions $u_k(x)$. Consider the initial value problem (for $x \geq 0$, with θ as defined above)

$$u''(x) + \lambda f(u(x)) = 0, \quad u(0) = \theta, \quad u'(0) = 0. \tag{3.123}$$

Since $f(\theta) > 0$, it follows that $u(x)$ is concave, and hence decreasing for small $x > 0$. Multiplying the equation (3.123) by u' and integrating, we conclude, in view of our initial conditions, that

$$\frac{1}{2}u'^2 + \lambda F(u) = 0. \tag{3.124}$$

The solution of (3.123) cannot decrease and be positive indefinitely. Indeed, if that was the case, $u(x)$ would have to converge to a constant $0 < c_0 < \theta$

as $x \to \infty$. We then obtain a contradiction from (3.124), since for large x, $u'(x)$ must be small, while the $\lambda F(u)$ term is negative and bounded away from zero.

It follows that either the solution of (3.123) becomes zero at some ξ, or else it stops being decreasing at some η, i.e., $u'(\eta) = 0$. In the first case, it follows from (3.124) that $u'(\xi) = 0$, and in the second case it follows from the same equation that $u(\eta) = 0$. In either case, we obtain a positive solution, with zero slope at its first root, i.e., $u(\xi) = u'(\xi) = 0$. Integrating (3.124),

$$\xi = \int_0^\theta \frac{du}{\sqrt{-2\lambda F(u)}}. \tag{3.125}$$

The value of $\xi = \frac{1}{2}$ corresponds to the first touchdown solution $u_1(x)$. Indeed, we have a non-negative solution with zero slope at $x = \frac{1}{2}$, and by symmetry, we have a non-negative solution on $(-\frac{1}{2}, \frac{1}{2})$, with zero slope at the end points. Translating this solution by $\frac{1}{2}$ to the right, we obtain $u_1(x)$. When $\xi = \frac{1}{2}$, we obtain from (3.125) the corresponding value of $\lambda_1 = 2 \left(\int_0^\theta \frac{du}{\sqrt{-F(u)}} \right)^2$. The k-th touchdown solution, $u_k(x)$, is obtained when $\xi = \frac{1}{2k}$, which happens at $\lambda = \lambda_k$, defined by (3.122). $\diamond$

Next, we show that solution curves stay bounded when λ ranges over a bounded interval, not including $\lambda = 0$.

Lemma 3.15. *Assume that $f(u)$ satisfies the conditions (3.113)-(3.115), and assume that $u_x(0, \lambda) \neq 0$ for $0 < l_1 \leq \lambda \leq l_2 < \infty$, with some constants l_1, l_2. Then for any solution branch $u(x, \lambda)$ of (3.112), there exists a constant c, such that*

$$|u(x, \lambda)| < c \text{ for all } \lambda \in [l_1, l_2]. \tag{3.126}$$

Proof: Assume, on the contrary, that $|u(x, \lambda)|$ becomes unbounded when $\lambda \to \lambda_0 \in [l_1, l_2]$. Assume that $u(x)$ is positive on some subinterval $I \equiv (a, b) \in (0, 1)$, and that $u(a) = u(b) = 0$. As we change λ, the roots of $u(x, \lambda)$ change, so that we can think of roots as continuous functions $a = a(\lambda)$, and $b = b(\lambda)$, and consider a variable interval $I = I_\lambda = (a(\lambda), b(\lambda))$. Writing our equation in the form

$$u'' + \lambda \frac{f(u)}{u} u = 0,$$

then using Sturm's comparison theorem, and our condition (3.114), we see that the length of the interval on which $|u(x, \lambda)|$ becomes large must be decreasing. I.e., given any $\epsilon > 0$ small, we can find a constant $M > 0$ and an interval $J \in I$ of length less than ϵ, and centered around the midpoint of I, so that

$$u(x, \lambda) < M \text{ for } x \in I \setminus J, \text{ and all } \lambda \in [l_1, l_2]. \qquad (3.127)$$

Since the "energy", $\frac{1}{2}u'^2(x) + \lambda F(u(x))$, is constant, we have, denoting $u_0 = u(\frac{a+b}{2})$,

$$u'^2(a) = 2\lambda \int_0^{u_0} f(u)\, du. \qquad (3.128)$$

Since $u_0 \to \infty$ as $\lambda \to \lambda_0$, it follows that $u'(a)$ must get large. Similarly, $u'(x)$ must get large for all $x \in I \setminus J$. We conclude that the interval I must be shrinking, since otherwise we get a contradiction with (3.127) (a bounded and increasing function cannot have unbounded derivatives on an interval, whose length is bounded from below).

If $u(x, \lambda)$ is a positive solution, we are done, since we may take $I = (0, 1)$, and this interval is not shrinking. Otherwise, we turn our attention to an adjacent to I interval, where the solution is negative, call it I_1. Denoting by $u_1 < 0$ the minimum value of $u(x, \lambda)$ on this interval, we see, using the energy, $\frac{1}{2}u'^2(x) + \lambda F(u(x)) = constant$, that $u_1 \to -\infty$ as $\lambda \to \lambda_0$. Using Sturm's comparison theorem, as before, we see that I_1 is also shrinking. Since by Lemma 3.11, the nodal structure of $u(x, \lambda)$ is preserved, while the length of all intervals of constant sign is decreasing, we obtain a contradiction. $\diamondsuit$

We describe, next, the structure of the solution set of (3.112) in a small neighborhood of a touchdown solution. If $u(x)$ is a touchdown solution, then $u'(x)$ spans the solution set of the corresponding linearized problem. The function $u'(x)$ is odd with respect to $x = \frac{1}{2}$, the mid-point of the interval, while the solution $u(x)$ is even with respect to $x = \frac{1}{2}$. If we restrict to the space of functions, which are even with respect to $x = \frac{1}{2}$, then the touchdown solution $u(x)$ is *non-singular*, and hence we can continue it for both increasing and decreasing λ, as a even with respect to $x = \frac{1}{2}$ solution. Denote by $U(x, \lambda)$ the curve of such symmetric solutions. For x close to the interior roots of the touchdown solution, solutions on $U(x, \lambda)$ will be negative for both $\lambda > \lambda_k$ and $\lambda < \lambda_k$, because points of positive local minimum are impossible. Near the end-points, solutions on $U(x, \lambda)$ are negative on one side of $\lambda = \lambda_k$, and positive on the other side, as we shall deduce by Lemma 3.13. Thus, we have two branches of symmetric solutions through (λ_k, u_k), one forward in λ, and one backward. We then consider $v(x, \lambda) = u(x, \lambda) - U(x, \lambda)$, and show that bifurcation from zero happens at $\lambda = \lambda_k$, giving us two branches of symmetry-breaking solutions. All four solution branches, emerging at (λ_k, u_k), will have to turn around and tend to infinity, as $\lambda \to 0$. We state next the main result of this section, based on P. Korman [84], which quantifies the above remarks.

Theorem 3.17. *Consider the problem (3.112), and assume that $f(u)$ satisfies the conditions (3.113), (3.114), and (3.115). Then, in addition to the numbers λ_k defined by (3.122), there exist a sequence of positive numbers $\{\mu_k\}$, with $\lambda_k < \mu_k$ for all k, so that for all $0 < \lambda < \lambda_k$, the problem (3.112) has at least two symmetric solutions with $2k$ interior zeros, and at least two asymmetric solutions with $2k - 1$ interior zeros, while for $\lambda > \mu_k$ only solutions with at least $2k + 1$ interior zeros may exist.*

Proof: We show that the pitchfork bifurcation occurs at each of the points (λ_k, u_k). Differentiating the equation (3.112), we see that any solution of the linearized problem (3.116) is a multiple of $u'_k(x)$.

We recast our equation (3.112) in the operator form $F(\lambda, u) = 0$, where $F : R \times C_0^2(0, 1) \to C(0, 1)$ is defined as follows ($C_0^2(0, 1)$ denotes functions in $C^2(0, 1) \cap C[0, 1]$, vanishing at $x = 0$, and $x = 1$)

$$F(\lambda, u) = u'' + \lambda f(u) = 0. \tag{3.129}$$

We denote by X the subspace of $C_0^2(0, 1)$, consisting of functions that are even with respect to $x = \frac{1}{2}$, and by Y the subspace of functions in $C_0^2(0, 1)$ that are odd with respect to $x = \frac{1}{2}$. Let Z be a subspace of $C(0, 1)$, consisting of functions that are even with respect to $x = \frac{1}{2}$. We now restrict our equation (3.129) to X, i.e., consider F as a map $R \times X \to Z$. We shall show that the equation (3.129) defines a curve on $R \times X$ (the boundary conditions, $u(0) = u(1) = 0$, are implicit here). The point $\lambda = \lambda_k$, $u = u_k$ lies on this curve. The linearized equation at this point

$$F_u(\lambda_k, u_k)w = w'' + \lambda_k f'(u_k)w = 0, \quad w(0) = w(1) = 0$$

has no nontrivial solutions in X, since its solution set is spanned by $u'_k(x) \notin X$. This implies that $F_u(\lambda_k, u_k)$ is an injection of X into Z. To see that it is also onto, take an arbitrary $g(x) \in Z$, and consider the problem

$$w'' + \lambda_k f'(u_k)w = g(x), \quad w(0) = w(1) = 0. \tag{3.130}$$

Since $\int_0^1 g u'_k \, dx = 0$, we see that the problem (3.130) is solvable. Writing its solution in the form $w = w_e + w_o$, with $w_e \in X$ and $w_o \in Y$, we see that w_e is also a solution of (3.130), and hence $F_u(\lambda_k, u_k)$ is onto (w_o is a constant multiple of u'_k). By the Implicit Function Theorem, we have a curve of symmetric solutions passing through (λ_k, u_k).

We claim, next, that solutions on this curve become negative near the end-points $x = 0, 1$, for $\lambda > \lambda_k$. This will follow from the fact that $u_\lambda(x, \lambda_k)$ is negative near $x = 0, 1$. Indeed, by Lemma 3.13,

$$u'_\lambda(0, \lambda_k) = \frac{1}{4} f(0) < 0, \tag{3.131}$$

and hence the symmetric solution becomes negative near the end-points, for increasing λ (because $u_\lambda(x, \lambda_k)$ is negative near the end-points).

We denote by $U = U(x, \lambda) = U(x)$, the curve of symmetric solutions passing through (λ_k, u_k) (to simplify the notation, we sometimes suppress the dependence on λ). Set $u(x) = U(x) + v(x)$. From the previous discussion, it follows that

$$U(x, \lambda_k) = u_k(x), \quad U_\lambda(x, \lambda_k) = u_\lambda(x, \lambda_k -). \tag{3.132}$$

We rewrite our equation (3.112) as

$$G(\lambda, v) \equiv v'' + \lambda \left[f(U + v) - f(U) \right] = 0, \quad v(0) = v(1) = 0. \tag{3.133}$$

The problem (3.133) has a trivial solution $v = 0$, for λ near λ_k. We are looking now for nontrivial solutions of (3.133), bifurcating off the trivial solution at the point $\lambda = \lambda_k$, $v = 0$. The linearized problem at this point

$$w'' + \lambda_k f'(u_k) w = 0, \quad w(0) = w(1) = 0, \tag{3.134}$$

has a one-dimensional null-space, spanned by u'_k. It is well known, by the standard elliptic theory, that the left hand side of (3.134) is a Fredholm operator of index zero, and so the range of $G_v(\lambda_k, 0)$ has codimension one. We shall show that the Crandall-Rabinowitz theorem on bifurcation from simple eigenvalues applies at this point (the Theorem 1.7 in [39]). In view of the above remarks, we only need to verify the crucial "transversality" condition: $G_{\lambda v}(\lambda_k, 0) u'_k \notin R(G_v(\lambda_k, 0))$. Since

$$G_{\lambda v}(\lambda_k, 0) u'_k = \lambda_k f''(u_k) u_\lambda u'_k + f'(u_k) u'_k,$$

we need to show, in view of the Fredholm alternative, that $G_{\lambda v}(\lambda_k, 0) u'_k$ is not orthogonal to the kernel of $G_v(\lambda_k, 0)$, i.e., $\lambda_k \int_0^1 f''(u_k) u_\lambda u_k'^2 \, dx + \int_0^1 f'(u_k) u_k'^2 \, dx \neq 0$. This will follow from a more precise inequality,

$$\lambda_k \int_0^1 f''(u_k) u_\lambda u_k'^2 \, dx + \int_0^1 f'(u_k) u_k'^2 \, dx > 0. \tag{3.135}$$

We proceed similarly to M. Ramaswamy [155] . To simplify the notation, we drop the subscript k in the intermediate steps, writing u instead of u_k and u' instead of u'_k. Integrating by parts, we express the first integral in (3.135)

$$\int_0^1 f''(u)u_\lambda u'^2 \, dx = \int_0^1 (f'(u))' \, u_\lambda u' \, dx \tag{3.136}$$
$$= -\int_0^1 f'(u)u'_\lambda u' \, dx - \int_0^1 f'(u)u_\lambda u'' \, dx.$$

Similarly, we rewrite the first term on the right in (3.136)

$$-\int_0^1 f'(u)u'_\lambda u' \, dx = \int_0^1 f(u)u''_\lambda \, dx + 2f(0)u'_\lambda(0),$$

and also the second term on the left in (3.135)

$$\int_0^1 f'(u)u'^2 \, dx \equiv \int_0^1 (f(u))' \, u' \, dx = -\int_0^1 fu'' \, dx = \lambda_k \int_0^1 f^2 \, dx.$$

Plugging these formulas into (3.135), and using the equations satisfied by u and u_λ respectively, we get

$$\lambda_k \int_0^1 f''(u_k)u_\lambda u'^2_k \, dx + \int_0^1 f'(u_k)u'^2_k \, dx = 2\lambda_k f(0)u'_\lambda(0).$$

In view of (3.131), we conclude (3.135).

Applying the Theorem 1.7 in [39], we conclude that in addition to $U(\lambda, x)$, there is another curve of solutions passing through (λ_k, u_k), having the form

$$\lambda = \lambda_k + \tau(s), \quad u = u_k + su'_k + z(s), \tag{3.137}$$

with the parameter s defined on some interval around $s = 0$, and $\tau(0) = \tau'(0) = 0$, $z(0) = z'(0) = 0$. This provides us with infinitely many points of pitchfork bifurcation. For $s > 0$ ($s < 0$), the solution $u(x)$ in (3.137) has $2k - 1$ interior zeros: two zeros near each root of $u_k(x)$, and one near $x = 1$ ($x = 0$).

We claim that no solution curve can pass through two points of pitchfork bifurcation. Indeed, non-symmetric solution curves passing through different u_k, have different number of zeros. Since the number of zeros is preserved on each solution curve, no such curve can pass through two different u_k. For the symmetric curves, the front curve at (λ_k, u_k) and the tail end at (λ_{k+1}, u_{k+1}) do have the same number of zeros, but they have the opposite order for regions of positivity and negativity, and hence cannot link up. Hence, the condition

$$u'(1, \lambda) \neq 0$$

is satisfied everywhere on each solution curve, except for a single point of pitchfork bifurcation. If the operator $F_u(\lambda, u)w$, defined previously, is invertible, the solution curve can be continued by the implicit function theorem. Otherwise, in view of Lemma 3.12, the Crandall-Rabinowitz bifurcation Theorem 1.2 applies, and hence all of the solution curves can be continued globally.

Next, we observe that no solution curve can continue for arbitrary large $\lambda > 0$, and hence all of the curves will have to turn around eventually, and tend to infinity as $\lambda \to 0$. Indeed, by Sturm's comparison theorem, the regions where solution is positive (negative) would have to shrink as $\lambda \to \infty$, but the nodal structure remains constant on each solution curve, which results in a contradiction.

Since by Lemma 3.15, solution curves cannot go to infinity at a positive λ, and all solutions disappear at $\lambda = 0$, it follows that all solution curves tend to infinity as $\lambda \to 0$ (the maximum value of solution is monotone on each solution curve). $\diamond$

Remark. An example of nonlinearity, for which the theorem applies, is furnished by $f(u) = u|u|^p - 1$, with any real $p > 0$. (For $0 < p < 1$, the function $f(u)$ is not of class C^2 at $u = 0$. However, examining the proof, we see that all of the integrals are still defined, and the proof does not need any changes.) On the other hand, the picture is different for e.g., $f(u) = u^2 - 1$. In this case, there are still infinitely many solution curves with pitchfork bifurcation, but this time $u = -1$ is a stable root, at which negative parts of solutions can accumulate, and in fact all solution curves continue as $\lambda \to \infty$, see H.P. McKean and J.C. Scovel [131], where a detailed study is given for this nonlinearity. Using elliptic functions, H.P. McKean and J.C. Scovel [131] showed that the solution set consists of infinitely many identically looking curves, with one point of pitchfork bifurcation on each curve. P. Korman [81] has extended their result, by using bifurcation analysis, similar to the above. We also mention that V. Anuradha and R. Shivaji [13] used time maps to get some related results.

3.9 An oscillatory bifurcation from zero: A model example

If one considers the problem

$$u'' + \lambda(u + g(u)) = 0, \quad -1 < x < 1, \quad u(-1) = u(1) = 0, \qquad (3.138)$$

with $\lim_{u \to 0} \frac{g(u)}{u} = 0$, then by the classical result of M.G. Crandall and P.H. Rabinowitz [39], a curve of positive solutions bifurcates from zero solution at $\lambda = \lambda_1 = \frac{\pi^2}{4}$, the principal Dirichlet eigenvalue of $-u''$ on $(-1, 1)$, corresponding to $\phi_1(x) = \cos \frac{\pi}{2} x$. It turns out that a similar situation may occur, when $g(u)$ is oscillatory for small u, instead of being small. Here we consider a model problem, and more general results can be found in P. Korman [97]. That paper also contains some results on oscillatory bifurcation from infinity, as well as similar results for balls in R^n, and for non-autonomous ODE's. Some related results can be found in A. Cañada [28].

We consider positive solutions (of class $C^2(0, 1) \cap C[0, 1]$) of a model problem

$$u'' + \lambda \left(u + \frac{1}{2} u \sin \frac{1}{u} \right) = 0, \quad -1 < x < 1, \quad u(-1) = u(1) = 0, \quad (3.139)$$

depending on a positive parameter λ. The classical result of M.G. Crandall and P.H. Rabinowitz [39] on bifurcation from zero does not apply here. However, we show that bifurcation from zero at λ_1 does indeed happen, and the solution curve makes infinitely many turns in any neighborhood of $(\lambda_1, 0)$, see the bifurcation diagram in Figure 3.6, computed by *Mathematica* (produced for the problem posed on the interval $x \in (0, 1)$, which can be reduced to (3.139) by scaling). Can one use this equation to model tornadoes?

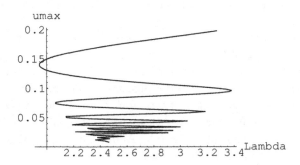

Fig. 3.6 Bifurcation diagram, near zero, for the problem (3.139)

Theorem 3.18. *There is a curve of positive solutions of the problem (3.139), bifurcating from infinity at $\lambda = \lambda_1$. The other side of this curve* *tends to zero, as* $\lambda \to \lambda_1$. *Moreover, in any neighborhood of* $(\lambda_1, 0)$, *this curve intersects the line* $\lambda = \lambda_1$ *infinitely many times, and the solution tends to a constant multiple of* $\phi_1(x)$, *as it approaches* $(\lambda_1, 0)$.

Proof: The proof is similar to that of Theorem 2.34 from the preceding chapter, so we shall be brief. By standard results on bifurcation from infinity, there is a curve of positive solutions, bifurcating from infinity at $\lambda = \lambda_1$. As before, we see that positive solutions of (3.139) are constrained to a strip

$$\frac{2}{3}\lambda_1 < \lambda < 2\lambda_1.$$

As before, we let $u = \alpha v$ in (3.139), with $\alpha = u(0)$, and $v(0) = 1$. Then (3.139) transforms to

$$v'' + \lambda\left(v + \frac{1}{2}v\sin\mu\frac{1}{v}\right) = 0, \quad -1 < x < 1, \quad v(-1) = v(1) = 0, \quad (3.140)$$

where $\mu = 1/\alpha$. Observe that $\mu \to \infty$, as $\alpha \to 0$ (i.e., when u tends to zero). Multiplying the equation (3.140) by ϕ_1, and integrating

$$(\lambda_1 - \lambda)\int_{-1}^{1} v(x)\phi_1(x)\,dx = \frac{1}{2}\lambda\int_{-1}^{1} v(x)\sin\mu\frac{1}{v(x)}\phi_1(x)\,dx. \quad (3.141)$$

Again, we observe that the integral on the right can be written as the imaginary part of a complex integral $\int_{-1}^{1} v(x)\phi_1(x)e^{i\mu\frac{1}{v(x)}}\,dx$, to which we apply the Lemma 2.25 (observe that the second derivative of $\frac{1}{v(x)}$ at zero is equal to $-v''(0)$), obtaining

$$\int_{-1}^{1} v(x)\phi_1(x)e^{i\mu\frac{1}{v(x)}}\,dx = e^{i\left(\mu - \frac{\pi}{4}\right)}\sqrt{\frac{\pi}{\mu|v''(0)|}} + O\left(\frac{1}{\mu}\right).$$

As before, we conclude from (3.141) that $\lambda_1 - \lambda$ changes sign infinitely many times, and it tends to zero, as $\mu \to \infty$. $\diamond$

3.10 Exact multiplicity for Hamiltonian systems

In case of space dimension equal to one, the Hamiltonian system, considered in the preceding chapter, takes the form (here $u = u(x)$ and $v = v(x)$)

$$u'' + \lambda H_v(u, v) = 0, \quad \text{for } -1 < x < 1, \quad u(-1) = u(1) = 0 \quad (3.142)$$

$$v'' + \lambda H_u(u, v) = 0, \quad \text{for } -1 < x < 1, \quad v(-1) = v(1) = 0.$$

We show that one can get considerably more detailed results in the one-dimensional case, including an exact multiplicity result.

Let us recall the conditions we have imposed on $H(u, v)$, and on the solution:

$$H_{vv} > 0, \text{ and } H_{uu} > 0 \text{ for all } (u, v) > 0, \tag{3.143}$$

$$H_{uv} \leq 0, \quad \text{for all } (u, v) \geq 0, \tag{3.144}$$

$$|u'(1)| + |v'(1)| > 0. \tag{3.145}$$

The corresponding linearized problem is

$$w'' + \lambda H_{vu}w + \lambda H_{vv}z = 0, \quad \text{for } -1 < x < 1, \quad w(-1) = w(1) = 0 \tag{3.146}$$
$$z'' + \lambda H_{uu}w + \lambda H_{uv}z = 0, \quad \text{for } -1 < x < 1, \quad z(-1) = z(1) = 0.$$

Theorem 3.19. *Assume that the conditions (3.143)-(3.145) hold. Assume also that*

$$H_v > 0, \quad \text{and } H_u > 0, \quad \text{for all } (u, v) > 0. \tag{3.147}$$

Then any non-trivial solution of (3.146) can be chosen to be positive on $(-1, 1)$ *(i.e., $w(x) > 0$ and $z(x) > 0$ on $(-1, 1)$).*

Proof: Recall that by W. Troy [184], $u(x)$ and $v(x)$ are both even functions, with $u'(x) < 0$ and $v'(x) < 0$ on $(0, 1)$. We claim that the functions $w(x)$ and $z(x)$ are also even. Indeed, assuming otherwise, $(w(-x), z(-x))$ would give us another solution of (3.146), linearly independent from $(w(x), z(x))$, contradicting that the null space is one dimensional, by Lemma 2.38. Differentiating the system (3.142), and denoting $p = u'$ and $q = v'$, we have

$$p'' + \lambda H_{vu}p + \lambda H_{vv}q = 0 \tag{3.148}$$
$$q'' + \lambda H_{uu}p + \lambda H_{uv}q = 0.$$

By our condition (3.147), $p' = u'' = -\lambda H_v < 0$, and $q' = v'' = -\lambda H_u < 0$. Multiply the first equation in (3.146) by q, and subtract from that the first equation in (3.148), multiplied by z,

$$w''q - p''z + \lambda H_{vu}wq - \lambda H_{vu}pz = 0.$$

Similarly, from the second equations of (3.146) and (3.148), we obtain

$$z''p - q''w + \lambda H_{uv}zp - \lambda H_{vu}wq = 0.$$

Adding the above two identities,

$$(w'q - p'z + z'p - q'w)' = 0,$$

i.e., $I(x) \equiv w'q - p'z + z'p - q'w = constant$. By Lemma 2.37, we may assume that

$$w'(1) < 0, \quad \text{and} \quad z'(1) < 0. \tag{3.149}$$

In view of (3.145), we then have

$$I(\alpha) = I(1) = w'(1)v'(1) + z'(1)u'(1) > 0. \tag{3.150}$$

By (3.149), it follows that $w(x)$ and $z(x)$ are both positive near $x = 1$. Assume, contrary to what we want to prove, that $w(x)$ vanishes on $(0, 1)$. Then $z(x)$ also vanishes on $(0, 1)$, since if $z(x)$ were positive, then w would have to be positive too, by the maximum principle. Let ξ denote the largest root of $w(x)$ on $(0, 1)$, and let α be the largest root of $z(x)$ on $(0, 1)$. Assume, first, that $\alpha < \xi$, i.e., $z(x) > 0$ on $(\alpha, 1)$. From the first equation in (3.146), we see that $w(x)$ cannot take negative minimums on $(\alpha, 1)$. This implies that (see Figure 3.7),

$$w(\alpha) < 0 \quad \text{and} \quad w'(\alpha) \geq 0. \tag{3.151}$$

We then have

$$I(\alpha) = w'(\alpha)q(\alpha) + z'(\alpha)p(\alpha) - q'(\alpha)w(\alpha) < 0, \tag{3.152}$$

since the first two terms on the left are non-positive, and the third one is negative, contradicting (3.150).

The case $\xi < \alpha$ is similar, while in the case $\xi = \alpha$, we have $w(\alpha) = 0$, and (3.152) changes to $I(\alpha) \leq 0$, which still results in a contradiction. $\diamond$

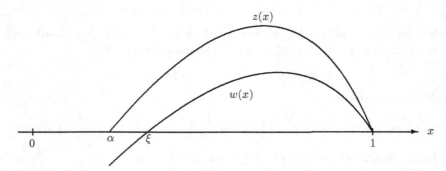

Fig. 3.7 The functions $w(x)$ and $z(x)$

We consider, next, the main special case when $H(u, v) = F(v) + G(u)$, and we denote $F'(v) = f(v)$, $G'(u) = g(u)$. The system (3.142) then becomes

$$u'' + \lambda f(v) = 0, \quad \text{for } -1 < x < 1, \quad u(-1) = u(1) = 0 \tag{3.153}$$

$$v'' + \lambda g(u) = 0, \quad \text{for } -1 < x < 1, \quad v(-1) = v(1) = 0.$$

The corresponding linearized problem takes the form

$$w'' + \lambda f'(v)z = 0, \quad \text{for } -1 < x < 1, \quad w(-1) = w(1) = 0 \quad (3.154)$$
$$z'' + \lambda g'(u)w = 0, \quad \text{for } -1 < x < 1, \quad z(-1) = z(1) = 0.$$

We have the following non-degeneracy result.

Theorem 3.20. *Assume that $f(t)$, $g(t) \in C^1([0, \infty))$ satisfy*

$$f(t) > 0, \quad f'(t) > 0, \quad g(t) > 0, \quad g'(t) > 0 \quad \text{for all } t > 0, \quad (3.155)$$

and in addition assume that

$$f'(t) \geq \frac{f(t)}{t}, \quad \text{and} \quad g'(t) \geq \frac{g(t)}{t} \quad \text{for all } t > 0, \quad (3.156)$$

with at least one of these inequalities being strict almost everywhere. Then any positive solution of (3.153) is non-singular, i.e., the linearized problem (3.154) admits only the trivial solution $w = z = 0$.

Proof: The positive functions $u(x)$ and $v(x)$ are concave, and so the condition (3.145) holds, and the Theorem 3.19 applies. Assume, on the contrary, that the linearized problem (3.154) has a non-trivial solution, with $w > 0$ and $z > 0$, in view of Theorem 3.19. We multiply the first equation in (3.153) by z, and subtract from that the first equation in (3.154), multiplied by v,

$$u''z + \lambda f(v)z - vw'' - \lambda f'(v)zv = 0.$$

Similarly, we multiply the second equation in (3.153) by w, and subtract from that the second equation in (3.154), multiplied by u,

$$v''w + \lambda g(u)w - uz'' - \lambda g'(u)uw = 0.$$

Adding the results,

$$(u'z - uz')' + (v'w - vw')' - \lambda \left[f'(v) - \frac{f(v)}{v} \right] vz - \lambda \left[g'(u) - \frac{g(u)}{u} \right] uw = 0.$$

Integrating this over $(-1, 1)$, we get a contradiction. $\diamond$

As an application, we consider the following case of a freely supported elastic beam ($p > 1$ is a constant)

$$u'''' = \lambda(u + u^p), \quad -1 < x < 1, \quad (3.157)$$
$$u(-1) = u''(-1) = u(1) = u''(1) = 0.$$

Let us denote by λ_1 the principal eigenvalue of u'''' on $(-1, 1)$, subject to the boundary conditions in (3.157).

Proposition 3.2. *The set of positive solutions of (3.157) consists of a single curve, bifurcating from the trivial solution at $\lambda = \lambda_1$. This curve continues without any turns for all $0 < \lambda < \lambda_1$. The maximum value of solution, $u(0, \lambda)$, is monotone increasing on this curve, and $\lim_{\lambda \to 0} u(0, \lambda) = \infty$. So that the problem (3.157) has a unique positive solution for $\lambda \in (0, \lambda_1)$, and no positive solution for $\lambda \geq \lambda_1$.*

Proof: By the standard result on bifurcation from zero, see M.G. Crandall and P.H. Rabinowitz [39], a curve of positive solutions bifurcates from zero at $\lambda = \lambda_1$, in the direction of decreasing λ. We put our problem in the system form (3.153)

$$u'' = \mu v, \quad \text{for } -1 < x < 1, \ u(-1) = u(1) = 0$$
$$v'' = \mu(u + u^p), \quad \text{for } -1 < x < 1, \ v(-1) = v(1) = 0,$$

where $\mu = \sqrt{\lambda}$. Theorem 3.20 applies here, and so the solution curve continues globally, without any turns. By P. Korman and J. Shi [119], maximum value of solution, $u(0, \lambda)$, is monotone increasing on this curve, and so $u(0, \lambda)$ tends to infinity. We claim that the solution curve goes to infinity as $\lambda \to 0$. This follows from the following a priori estimate: if λ belongs to a compact subinterval of $(0, \infty)$, then there exists a constant $M > 0$, such that any solution of (3.157) satisfies

$$|u|_{C^4(-1,1)} \leq M.$$

The proof of this estimate is almost identical to that of Lemma 3.26 of the next section, so we omit it (the proof is even a little easier here, since here $u(x)$ is concave). Finally, since this solution curve "takes up" all possible values of $u(0)$, from zero to infinity, no other positive solutions are possible.
$\diamond$

The following exact multiplicity result was proved in P. Korman [87].

Theorem 3.21. *Assume that the functions $f(t)$, $g(t) \in C^2(\bar{R}_+)$ satisfy the conditions in (3.155), and in addition,*

$$f(0) + g(0) > 0, \tag{3.158}$$

$$f''(t) \geq 0, \quad \text{and} \quad g''(t) \geq 0 \quad \text{for all } t > 0, \tag{3.159}$$

with at least one of the inequalities being strict on a set of positive measure,

$$f(t) > \alpha t, \quad \text{and} \quad g(t) > \alpha t, \quad \text{for some } \alpha > 0, \text{ and all } t > 0, \tag{3.160}$$

$$\text{either} \quad \lim_{t \to \infty} \frac{f(t)}{t} = \infty, \quad \text{or} \quad \lim_{t \to \infty} \frac{g(t)}{t} = \infty. \tag{3.161}$$

Then there is a critical $\lambda_0 > 0$, such that the problem (3.153) has exactly 2, 1 or 0 positive solutions, depending on whether $\lambda < \lambda_0$, $\lambda = \lambda_0$ or $\lambda > \lambda_0$. Moreover, all solutions lie on a single smooth solution curve, which, for $\lambda < \lambda_0$, has two branches denoted by $0 < (u^-(x, \lambda), v^-(x, \lambda)) < (u^+(x, \lambda), v^+(x, \lambda))$ (componentwise), with $(u^-(x, \lambda), v^-(x, \lambda))$ strictly monotone increasing (componentwise) in λ. We also have $\lim_{\lambda \to 0}(u^-(x, \lambda), v^-(x, \lambda)) = (0, 0)$, and $\lim_{\lambda \to 0}(u^+(x, \lambda), v^+(x, \lambda)) = (\infty, \infty)$ for all $x \in (-1, 1)$.

Proof: The proof follows a familiar pattern. By the Implicit Function Theorem, we can continue the trivial solution $u = v = 0$ at $\lambda = 0$, to conclude existence of positive solutions for small $\lambda > 0$.

Next, we show that no positive solution exists for λ large. We multiply each equation in (3.153) by $\cos \frac{\pi}{2}x$, integrate and add the results. Using the assumption (3.160), we conclude

$$-\frac{\pi^2}{4} \int_{-1}^{1} (u(x) + v(x)) \cos \frac{\pi}{2}x \, dx + \lambda\alpha \int_{-1}^{1} (u(x) + v(x)) \cos \frac{\pi}{2}x \, dx < 0,$$

which implies that $\lambda < \frac{\pi^2}{4\alpha}$.

Our next step is to show that a positive solution of (3.153) cannot become unbounded at any $\bar{\lambda} > 0$. The idea of the proof is that solution can become large only on a small interval, which is inconsistent with it being a concave function. Assume, on the contrary, that either $u(x)$ or $v(x)$ becomes unbounded as $\lambda \to \bar{\lambda}$, say $u(x)$. Since $u(x)$ is concave, it follows that it has to become large on any proper subinterval of $(-1, 1)$, say on $(0, 1/2)$. Using the assumption (3.160), it easily follows that $v(x)$ must also become unbounded on $(0, 1/2)$ (by using the Green's function representation, for example). Assume, for definiteness, that the first condition in (3.161) holds. Then for any $M > 0$, we have

$$f(v) > Mv \quad \text{on } (0, 1/2), \tag{3.162}$$

provided that λ is close to $\bar{\lambda}$. We now multiply the first equation in (3.153) by $\sin 2\pi x$, and integrate over $(0, 1/2)$. Using (3.162), we have (dropping two positive terms on the left)

$$-4\pi^2 \int_{0}^{\frac{1}{2}} u(x) \sin 2\pi x \, dx + \lambda M \int_{0}^{\frac{1}{2}} v(x) \sin 2\pi x \, dx < 0. \tag{3.163}$$

Similarly, from the second equation in (3.153),

$$-4\pi^2 \int_{0}^{\frac{1}{2}} v(x) \sin 2\pi x \, dx + \lambda\alpha \int_{0}^{\frac{1}{2}} u(x) \sin 2\pi x \, dx < 0. \tag{3.164}$$

Since M is large, the inequalities (3.163) and (3.164) contradict each other.

We now consider the positive solution of (3.153) at some small λ. If the corresponding linearized problem (3.154) has only the trivial solution ($w = 0, z = 0$), then by the Implicit Function Theorem, we can locally continue this solution for increasing λ. Since for large λ the problem (3.153) has no positive solutions, we cannot continue the curve indefinitely, with λ increasing. Let λ_0 be the supremum of λ's, for which we can continue the solution curve to the right. By above, solutions on this curve remain bounded as $\lambda \to \lambda_0$. Using elliptic estimates, we pass to the limit, and conclude that at λ_0, the problem (3.153) has a solution $(u_0, v_0) \in C^2[-1, 1] \times C^2[-1, 1]$. By the definition of λ_0, it follows that (u_0, v_0) is a singular solution, i.e., the problem (3.154) has a non-trivial solution, (w, z). By the Theorem 3.19, we have

$$w(x) > 0, \quad \text{and} \quad z(x) > 0 \quad \text{for all } x \in (-1, 1). \tag{3.165}$$

As we had explained in the preceding chapter, the Crandall-Rabinowitz Theorem 1.7 applies at (u_0, v_0), and hence we can continue the solution curve through this point. We now compute the direction of turn at (u_0, v_0) (and at any other singular point).

Near (λ_0, u_0, v_0), we represent the solution curve as $\lambda = \lambda(s)$, $(u, v) = (u(s), v(s))$, with $\lambda(0) = \lambda_0$, and $(u(0), v(0)) = (u_0, v_0)$. Notice that $\lambda'(0) = 0$, since the solution curve does not extend beyond λ_0. Also, by the Crandall-Rabinowitz Theorem 1.7, $(u_s(0), v_s(0)) = (w, z)$. Differentiating the first equation in (3.153) twice in s, and then letting $s = 0$, we obtain

$$u_{ss}'' + \lambda_0 f' v_{ss} + \lambda_0 f'' z^2 + \lambda''(0) f = 0, \quad u_{ss}(-1) = u_{ss}(1) = 0. \tag{3.166}$$

Similarly, from the second equation in (3.153), we obtain

$$v_{ss}'' + \lambda_0 g' u_{ss} + \lambda_0 g'' w^2 + \lambda''(0) g = 0, \quad v_{ss}(-1) = v_{ss}(1) = 0. \tag{3.167}$$

We now multiply (3.166) by z, the first equation in (3.154) by v_{ss}, subtract and integrate

$$- \int_{-1}^{1} z' u_{ss}' \, dx + \int_{-1}^{1} w' v_{ss}' \, dx + \lambda_0 \int_{-1}^{1} f''(v_0) z^3 \, dx \tag{3.168}$$
$$+ \lambda''(0) \int_{-1}^{1} f(v_0) z \, dx = 0.$$

Similarly, using (3.167) and the second equation in (3.154), we obtain

$$- \int_{-1}^{1} w' v_{ss}' \, dx + \int_{-1}^{1} z' u_{ss}' \, dx + \lambda_0 \int_{-1}^{1} g''(u_0) w^3 \, dx \tag{3.169}$$
$$+ \lambda''(0) \int_{-1}^{1} g(u_0) w \, dx = 0.$$

Adding (3.168) and (3.169), we have, using (3.159) and (3.165),

$$\lambda''(0) = -\lambda_0 \frac{\int_{-1}^{1} f''(v_0)z^3 \, dx + \int_{-1}^{1} g''(u_0)w^3 \, dx}{\int_{-1}^{1} f(v_0)z \, dx + \int_{-1}^{1} g(u_0)w \, dx} < 0. \qquad (3.170)$$

It follows that only "turns to the left" are possible at any singular solution of (3.153).

We see that after turning to the left at (λ_0, u_0, v_0), the solution curve continues to travel to the left, since turns to the right are impossible. By the result of P. Korman and J. Shi [119] (reviewed in the preceding chapter), we see that the maximum values $(u(0), v(0))$ keep increasing (componentwise), as we follow the curve for decreasing λ (after the turn). Since solutions cannot become unbounded at any $\lambda > 0$, and there are no non-trivial solutions at $\lambda = 0$, it follows that solutions on the curve become unbounded as $\lambda \downarrow 0$. $\diamond$

In the special case, when $f(v) = v$, we obtain an exact multiplicity result for the fourth-order equation, which we state explicitly next, given its physical significance. This model describes deflections of a freely supported elastic beam.

Theorem 3.22. *Consider the problem*

$$u'''' = \lambda f(u) \ \ on \ (-1, 1), \ \ u(-1) = u(1) = u''(-1) = u''(1) = 0. \quad (3.171)$$

Assume that $f(t) \in C^2(\bar{R}_+)$ satisfies the following conditions

$$f(0) > 0,$$

$$f'(t) > 0, \quad for \ all \ t > 0,$$

$$f''(t) \geq 0, \quad for \ all \ t > 0,$$

with the inequality being strict on a set of positive measure,

$$f(t) > \alpha t, \quad for \ some \ \alpha > 0, \ and \ all \ t > 0,$$

$$\lim_{t \to \infty} \frac{f(t)}{t} = \infty.$$

Then there is a critical $\lambda_0 > 0$ such that the problem (3.171) has exactly 2, 1 or 0 positive solutions, depending on whether $\lambda < \lambda_0$, $\lambda = \lambda_0$ or $\lambda > \lambda_0$. Moreover, all solutions lie on a unique smooth solution curve, which begins at $(\lambda = 0, u = 0)$, it makes a turn to the left at some $\lambda = \lambda_0$, and it tends to infinity for all $x \in (-1, 1)$, as $\lambda \to 0$.

The main example is $f(u) = e^u$.

3.11 Clamped elastic beam equation

Following P. Korman [91], we study positive solutions of the problem

$$u''''(x) = \lambda f(u(x)), \quad \text{for } x \in (0,1) \tag{3.172}$$
$$u(0) = u'(0) = u(1) = u'(1) = 0,$$

which describes the deflection $u(x)$ of an elastic beam, subjected to a non-linear force $f(u)$, and *clamped* at the end points. Here λ is a positive parameter, and $f(u) \in C^2(\bar{R}_+)$ is a given function. We are interested in existence and exact multiplicity of positive solutions. While similar second order problems have been extensively studied, relatively little is known about the problem (3.172). The likely reason is that fewer techniques are available for the fourth order problems. In particular, the strong maximum principle does not hold here. Also, the popular time map method appears to be completely inapplicable. Previous works on the problem include R. Dalmasso [41] and [42], D.R. Dunninger [52], and P. Korman [77]. These papers used shooting techniques, the Leray-Schauder degree, and monotone iterations. We use the bifurcation approach of this book, to obtain an exact multiplicity result for a class of convex $f(u)$, and a uniqueness result for a class of superlinear $f(u)$.

3.11.1 *Preliminary results*

We begin by studying a linear boundary value problem

$$v''''(x) = c(x)v(x), \quad c(x) > 0, \quad x \in (0,1) \tag{3.173}$$
$$v(0) = v'(0) = v(1) = v'(1) = 0,$$

where $c(x) \in C[0,1]$ is a given positive function.

Lemma 3.16. *If $v(x)$ solves (3.173), and $v''(0) = 0$, then $v(x) \equiv 0$.*

Proof: Assume, on the contrary, that $v(x)$ is a non-trivial solution, with $v''(0) = 0$. Consider $v'''(0)$. If $v'''(0) = 0$, then $v(x) \equiv 0$ by uniqueness for initial value problems. Otherwise, by the linearity of the problem, we may assume that $v'''(0) > 0$, and then $v(x)$ is positive for small x. We may assume that

$$v(x) > 0 \text{ for } x \in (0,\delta), \quad v(\delta) = 0, \quad v'(\delta) \leq 0, \tag{3.174}$$

where $\delta \in (0,1]$ is the supremum of the x's, for which $v(x) > 0$. Since from the equation (3.173), the function $v'''(x)$ is increasing, it follows that

$$v'''(x) > 0 \text{ for } x \in [0,\delta].$$

Hence, the function $v''(x)$ is increasing. Combining that with $v''(0) = 0$, we conclude that

$$v''(x) > 0 \quad \text{for } x \in (0, \delta].$$

It follows that the function $v'(x)$ is increasing. Combining that with $v'(0) = 0$, we conclude that

$$v'(x) > 0 \quad \text{for } x \in (0, \delta].$$

If $\delta = 1$, then the last inequality contradicts the boundary condition $v'(1) = 0$, otherwise we conclude that $v(\delta) > 0$, contradicting (3.174). $\diamondsuit$

Corollary The linear space of non-trivial solutions of (3.173) is either empty, or one-dimensional. Indeed, in view of the lemma, the set of solutions of (3.173) can be parameterized by the value of $v''(0)$.

A similar result holds for the nonlinear problem (3.172).

Lemma 3.17. *Assume that $f(u) \geq 0$ for $u > 0$. If $u(x)$ is a positive solution of (3.172), then*

$$u''(0) > 0, \quad \text{and } u''(1) > 0. \tag{3.175}$$

Proof: Since $u(x)$ is positive, $u''(0) \geq 0$. Assume, on the contrary, that $u''(0) = 0$. We consider $u'''(0) \equiv \alpha$. Clearly, $\alpha \geq 0$. Integrating the equation (3.172), we express

$$u(x) = \int_0^x \frac{(x - \xi)^3}{3!} f(u(\xi)) \, d\xi + \alpha \frac{x^3}{3!}.$$

But then $u(1) > 0$, contradicting the boundary condition $u(1) = 0$. $\diamondsuit$

The linearized problem for the problem (3.172) is

$$w''''(x) = \lambda f'(u)w, \ x \in (0,1), \ w(0) = w'(0) = w(1) = w'(1) = 0. \tag{3.176}$$

Lemma 3.18. *If $u(x)$ and $w(x)$ are solutions of the problems (3.172) and (3.176), respectively, then*

$$w'''u' - wu'''' - w''u'' + w'u''' = -u''(1)w''(1), \quad \text{for all } x \in [0,1]. \tag{3.177}$$

Proof: Differentiating the function $w'''u' - wu'''' - w''u'' + w'u'''$, and using the equations (3.172) and (3.176), we see that this function is constant on $[0,1]$. Evaluating this function at $x = 1$, gives us (3.177). $\diamondsuit$

We shall need the following lemma, to verify the hypothesis of the Crandall-Rabinowitz Theorem 1.7.

Lemma 3.19. *If $u(x)$ and $w(x)$ are solutions of the problems (3.172) and (3.176), respectively, then*

$$\int_0^1 f(u(x))w(x)\,dx = \frac{1}{4\lambda}u''(1)w''(1)\,. \qquad (3.178)$$

Proof: Integrate the identity (3.177) over $(0,1)$. Then integrate by parts in the first, third, and the fourth terms on the left, shifting all of the derivatives on u. All boundary terms vanish, in view of our boundary conditions. Then expressing u'''' from the equation (3.172), we conclude the lemma. $\diamond$

The following lemma provides an "energy" functional for the problem (3.172). As usual, we denote $F(u) = \int_0^u f(t)\,dt$.

Lemma 3.20. *If $u(x)$ is any solution of (3.172), then*

$$u'u''' - \frac{1}{2}u''^2 - \lambda F(u) = const, \quad for \ all \ x \in [0,1]\,. \qquad (3.179)$$

Proof: Just differentiate $u'u''' - \frac{1}{2}u''^2 - \lambda F(u)$, and see that this derivative is zero, by using the equation (3.172). $\diamond$

The following two lemmas describe the profile of positive solutions.

Lemma 3.21. *Assume that $f(u) > 0$ for $u > 0$. Then any positive solution of (3.172) has exactly one local (and hence global) maximum.*

Proof: Since $(u'')'' = f(u) > 0$, it follows that the function u'' is negative between any two of its consecutive roots. I.e., $u(x)$ is a concave function between any two consecutive points of inflection. Hence, the positive function $u(x)$ cannot have any points of local minimum, and the proof follows.

$\diamond$

Corollary The positive solution $u(x)$ has exactly two points of inflection. Indeed, by Lemma 3.17, $u(x)$ is convex near the end-points, so that it has at least two points of inflection, and by the proof of Lemma 3.21, there are at most two points of inflection.

The following lemma gives conditions for any positive solution of (3.172) to be symmetric. A more general symmetry result (in case of $f = f(x, u, u'')$) was given previously by R. Dalmasso [41]. For the particular case of (3.172), our proof of symmetry is much easier. Moreover, a similar argument is used in a lemma to follow.

Lemma 3.22. *For the problem (3.172), assume that $f(u) \subset C^1(0, \infty) \cap C[0, \infty)$ satisfies $f(u) > 0$, and $f'(u) \geq 0$, with the last inequality being strict almost everywhere. Then any positive solution of (3.172) is symmetric with respect to $x = \frac{1}{2}$. Moreover, $u'(x) > 0$ on $(0, \frac{1}{2})$.*

Proof: By Lemma 3.21, any positive solution $u(x)$ is unimodular, with a unique point of maximum, which we denote by x_0. Assuming, on the contrary, that $u(x)$ is not symmetric, we observe that $v(x) \equiv u(1-x)$ is then a different solution of (3.172). We have $u(0) = v(0) = 0$, $u'(0) = v'(0) = 0$, and by Lemma 3.20, $u''(0) = v''(0)$. By the uniqueness for the initial value problems, the third derivatives must be different at 0, so that we may assume, for definiteness, that $v'''(0) > u'''(0)$. It follows that $v(x) > u(x)$, for small x. Let ξ denote the first point where the graphs of $u(x)$ and $v(x)$ intersect, and if they never intersect, we let ξ be the point of maximum of $v(x)$ (i.e., $\xi = 1 - x_0$). Let $w(x) = v(x) - u(x)$. Then $w(x)$ satisfies

$$w'(\xi) \leq 0, \tag{3.180}$$

and

$$w'''' = c(x)w \quad \text{on } (0, \xi), \quad w(0) = w'(0) = w''(0) = 0, \quad w'''(0) > 0,$$

where $c(x) = \lambda \int_0^1 f'(\theta u + (1 - \theta)v)\, d\theta > 0$. Since $w(x)$ is positive for small x, it follows that $w'''(x)$ is increasing, and hence positive. The same is true for $w''(x)$, and then for $w'(x)$, for all $x \in (0, \xi)$. In particular, $w'(\xi) > 0$, contradicting (3.180).

We now turn to the last claim. Since $u(x)$ has only one point of inflection on $(0, 1/2)$, call it ξ, it follows that $u'(x)$ is increasing on $(0, \xi)$, decreasing on $(\xi, 1/2)$, while $u'(0) = u'(1/2) = 0$. It follows that $u'(x) > 0$, for all $x \in (0, 1/2)$. $\diamond$

By the last lemma, any positive solution of (3.172) has a global maximum at $x = \frac{1}{2}$. It turns out that this maximal value uniquely identifies the solution pair $(\lambda, u(x))$, similarly to the second order problems.

Lemma 3.23. *Assume that the conditions of Lemma 3.22 hold. Then the positive solutions of (3.172) are globally parameterized by their maximum values $u(\frac{1}{2}, \lambda)$. I.e., for every $p > 0$ there is at most one $\lambda > 0$, and at most one solution $u(x, \lambda)$ of (3.172), for which $u(\frac{1}{2}, \lambda) = p$.*

Proof: By shifting and then stretching of the x variable, we can replace the problem (3.172) by

$$u''''(x) = \lambda f(u(x)), \quad x \in (-1, 1) \tag{3.181}$$

$$u(-1) = u'(-1) = u(1) = u'(1) = 0,$$

with the maximum value $u(0, \lambda) = p$. By the preceding lemma, $u(x)$ is an even function. Then, $v(x) = u(\frac{1}{\sqrt[4]{\lambda}}x, \lambda)$ satisfies

$$v'''' = f(v) \text{ on } (0, \sqrt[4]{\lambda}), \qquad (3.182)$$
$$v(0) = p, \quad v'(0) = v'''(0) = 0, \quad v(\sqrt[4]{\lambda}) = v'(\sqrt[4]{\lambda}) = 0.$$

Assume there is a different solution $u_1(x, \mu)$ of (3.181), with $u_1(0, \mu) = p$. Consider, first, the case $\mu \neq \lambda$. Then $v_1(x) = u_1(\frac{1}{\sqrt[4]{\mu}}x, \mu)$ is another solution of the equation in (3.182), which has its first root at $\sqrt[4]{\mu}$. Since $v(x)$ and $v_1(x)$ are different solutions, by the uniqueness for initial value problems, we conclude that $v''(0) \neq v_1''(0)$, and we may assume that $v''(0) > v_1''(0)$. It follows that $v(x) > v_1(x)$, for $|x|$ small. Then $w(x) = v(x) - v_1(x)$ satisfies

$$w'''' = c(x)w \text{ for } x > 0, \qquad (3.183)$$
$$w(0) = w'(0) = w'''(0) = 0, \quad w''(0) > 0,$$

where $c(x) = \lambda \int_0^1 f'(\theta v + (1 - \theta)v_1) \, d\theta > 0$. We see from (3.183) that $w(x)$ and its first three derivatives are positive, for $x > 0$. On the other hand, either the functions $v(x)$ and $v_1(x)$ intersect at some $\xi > 0$, where $w'(\xi) \leq 0$, or else the smaller solution $v_1(x)$ reaches its root before $v(x)$ does, so that $w'(\sqrt[4]{\mu}) \leq 0$. In both cases, we have a contradiction (by above, $w'(x) > 0$).

In case $\mu = \lambda$, we argue similarly. In that case, either $v(x)$ and $v_1(x)$ intersect at some point ξ, where we have $w'(\xi) \leq 0$, or else these functions reach their first root at $\sqrt[4]{\lambda}$, where $w'(\sqrt[4]{\lambda}) = 0$, leading to the same contradiction as before. $\diamond$

The next lemma shows that solutions of the linearized problem cannot have two changes of monotonicity without changing the sign. We state it in a little more general form.

Lemma 3.24. *Let $a(x)$ be a positive continuous function. Then any nontrivial solution of the problem*

$$w'''' = a(x)w, \quad w(0) = w(1) = 0$$

cannot have points of non-negative local minimums, and points of nonpositive local maximums. In particular, under the conditions of Lemma 3.22, any non-trivial solution $w(x)$ of the linearized problem (3.176) cannot have non-negative local minimums, and non-positive local maximums.

Proof: Since $w(x)$ vanishes at the end-points of the interval $(0, 1)$, for any point of non-negative local minimum x_0, we can find two inflection points of $w(x)$, call them α and β, with $\alpha < x_0 < \beta$. On (α, β), $(w'')'' > 0$, which together with $w''(\alpha) = w''(\beta) = 0$, implies that $w(x)$ is concave, in particular $w''(x_0) < 0$, a contradiction. The other claim is proved similarly ($w(x)$ would have to be convex near the point of maximum). $\diamond$

We shall need the following lemma, which is a special case of the Theorem 3.1 of the classical paper of W. Leighton and Z. Nehari [124]. We present its proof for completeness.

Lemma 3.25. *Let* $a(x)$ *be a positive continuous function. Let* $u(x)$ *and* $v(x)$ *be non-trivial solutions of*

$$v''''(x) = a(x)v(x), \quad a(x) > 0 \quad x \in (0, 1), \tag{3.184}$$

which are not multiples of one another, such that $u(0) = v(0) = u(1) = v(1) = 0$. *Then the zeros of* $u(x)$ *and* $v(x)$ *in* $(0, 1)$ *separate each other.*

Proof: Assume, on the contrary, that $0 < \alpha < \beta < 1$ are two consecutive roots of $u(x)$, i.e., $u(\alpha) = u(\beta) = 0$, while $v(x) > 0$ on $[\alpha, \beta]$. We claim that $p(x) \equiv u'v - uv'$ has to vanish on (α, β). Indeed, assuming that, say, $p(x) > 0$ on (α, β), we integrate the identity $\left(\dfrac{u}{v}\right)' = \dfrac{p}{v^2} > 0$, obtaining

$$0 = \frac{u(\beta)}{v(\beta)} - \frac{u(\alpha)}{v(\alpha)} = \int_\alpha^\beta \frac{p}{v^2} \, dx > 0,$$

a contradiction, proving the claim. So let $p(x_0) = 0$ at some $x_0 \in (\alpha, \beta)$. We may regard $p(x_0)$ as a 2×2 Wronskian determinant. Since it vanishes at x_0, its columns are proportional, i.e.,

$$\begin{pmatrix} u(x_0) \\ u'(x_0) \end{pmatrix} = \gamma \begin{pmatrix} v(x_0) \\ v'(x_0) \end{pmatrix}, \tag{3.185}$$

with some constant γ. We now consider $w(x) \equiv u(x) - \gamma v(x)$, which is also a solution of (3.184), satisfying $w(0) = w(1) = 0$. In view of (3.185), $w(x)$ has a double root at x_0, i.e., $w(x_0) = w'(x_0) = 0$. If $w''(x_0) > 0 \; (< 0)$, then x_0 is a point of non-negative minimum (non-positive maximum), which is impossible by Lemma 3.24. Hence, $w''(x_0) = 0$. We now consider $w'''(x_0)$. If $w'''(x_0) = 0$, then $w(x) \equiv 0$, contradicting $w(\alpha) \neq 0$. Hence, we may assume that $w'''(x_0) > 0$. Then $w(x)$ is positive and convex to the right of x_0. But $w(1) = 0$. Hence, we can find an inflection point $\delta \in (x_0, 1)$ where $w(x)$ becomes concave, i.e., $w'''(\delta) \leq 0$. But, from the equation (3.184), $w'''(x)$ is increasing on (x_0, δ), a contradiction. $\diamond$

The following theorem is crucial for our uniqueness and exact multiplicity results.

Theorem 3.23. *Assume that the conditions of Lemma 3.22 hold. Then any non-trivial solution of the linearized problem (3.176) cannot vanish inside $(0, 1)$. I.e., we may assume that $w(x) > 0$ on $(0, 1)$.*

Proof: By Lemma 3.24, it suffices to prove that $w(x)$ is non-negative, i.e., that it cannot change sign. Assuming the contrary, let us ignore any roots of $w(x)$ where it "touches" the x-axis, and consider only the roots where $w(x)$ changes sign. We claim that $w(x)$ cannot change sign an odd number of times. If the contrary is true, we may assume, for definiteness, that $w(x)$ is negative near $x = 0$, and positive near $x = 1$. In view of Lemma 3.16, we then have $w''(0) < 0$ and $w''(1) > 0$. Setting $x = 0$ in (3.177), we have

$$u''(0)w''(0) = u''(1)w''(1).$$

This gives us a contradiction, since in view of Lemma 3.17, the quantity on the left is negative, while the one on the right is positive.

So, we may assume that $w(x)$ changes sign an even number of times. We claim that $w(x)$ is then even with respect to $x = \frac{1}{2}$. Indeed, since, by Lemma 3.22, $u(x)$ is even with respect to $x = \frac{1}{2}$, it follows that $w(1 - x)$ is a solution of the linearized problem (3.176). Since, by Lemma 3.16, the null-space of (3.176) is one-dimensional, we have

$$w(1 - x) = cw(x), \tag{3.186}$$

with a constant $c \neq 1$. (If $c = 1$, $w(x)$ is even, and the claim is proved.) It follows from (3.186) that $w(\frac{1}{2}) = 0$. We claim that $w'(\frac{1}{2}) \neq 0$. Indeed, assume, on the contrary, that $w'(\frac{1}{2}) = 0$. From (3.186), we conclude that $w''(\frac{1}{2}) = 0$. Then evaluating (3.177) at $x = 1/2$, we obtain $0 = u''(1)w''(1)$, which is a contradiction, since the quantity on the right is non-zero. Hence, $w'(\frac{1}{2}) \neq 0$, i.e., $w(x)$ changes sign at $x = \frac{1}{2}$. It follows that the number of sign changes for $w(x)$ is different on the intervals $(0, \frac{1}{2})$ and $(\frac{1}{2}, 1)$, which makes (3.186) impossible. Hence, $w(x)$ is even.

We show, next, that $w(x)$ cannot change sign twice. Assuming the contrary, we shall use that $w(x)$ is even with respect to $x = \frac{1}{2}$. Assuming, for definiteness, that $w(x)$ is positive near $x = 1$, we have $w''(1) > 0$, and $w(\frac{1}{2}) < 0$, $w'(\frac{1}{2}) = 0$, $w''(\frac{1}{2}) \geq 0$. (The last inequality follows by Lemma 3.24.) Then, from (3.177), we conclude

$$-w\left(\frac{1}{2}\right) f\left(u\left(\frac{1}{2}\right)\right) - u''\left(\frac{1}{2}\right) w''\left(\frac{1}{2}\right) = -u''(1)w''(1). \tag{3.187}$$

The first term on the left in (3.187) is positive, the second one is non-negative, while the right hand side is negative, a contradiction.

We conclude from the above discussion that $w(x)$ has to change sign at least four times, and it is an even function. Hence, $w(x)$ has at least two roots on $(0, \frac{1}{2})$. Differentiating the equation (3.172), we conclude that $u'(x)$ is a solution of the linearized equation (3.176), different from $w(x)$ ($u'(x)$ is odd, while $w(x)$ is even). Also, $u'(0) = w(0) = u'(1) = w(1) = 0$. By the Lemma 3.25 of W. Leighton and Z. Nehari [124], the roots of $w(x)$ and $u'(x)$ separate each other. This implies that $u'(x)$ must vanish inside $(0, \frac{1}{2})$, which is impossible by Lemma 3.22. We conclude that $w(x)$ cannot change sign, and hence it is non-negative on $(0, 1)$. By Lemma 3.24, $w(x)$ is actually positive. $\diamondsuit$

We shall need an a priori estimate, which was proved in D.R. Dunninger [52]. For completeness, we present a simpler proof, although we retain an idea from [52]. Notice that conditions of the next lemma do not imply symmetry of solutions.

Lemma 3.26. *Assume that $f(u) > 0$ for $u > 0$, and $\lim_{u \to \infty} \frac{f(u)}{u} = \infty$, while λ belongs to a compact interval $I \subset (0, \infty)$. Then there exists a constant $M > 0$, such that any solution of (3.172) satisfies*

$$|u|_{C^4(0,1)} \leq M. \tag{3.188}$$

Proof: We claim that there is a constant $m > 0$ such that any positive solution of (3.172) satisfies

$$u''(0) \leq m. \tag{3.189}$$

Assuming the claim for the moment, we now complete the proof. Evaluating the "energy" identity (3.179) at $x = 0$, we obtain

$$u'u''' - \frac{1}{2}u''^2 - \lambda F(u) = -\frac{1}{2}u''^2(0), \quad \text{for all } x \in [0, 1]. \tag{3.190}$$

Observe that $F(u) > 0$ for $u > 0$. Next, integrate (3.190) over $(0, 1)$. After integration by parts, we have

$$\frac{3}{2} \int_0^1 u''^2 \, dx \leq \frac{1}{2}m^2. \tag{3.191}$$

By the Sobolev embedding theorem, $|u|_{C^1(0,1)}$ is bounded, and then, using the equation (3.172), we obtain a bound on $|u|_{C^4(0,1)}$, concluding the lemma.

We now turn to proving the claim (3.189). Let $\mu_1 > 0$ and $\phi_1(x) > 0$ be the principal eigenpair of the eigenvalue problem

$$\phi'''' = \mu\phi \text{ in } (0,1), \quad \phi(0) = \phi'(0) = \phi(1) = \phi'(1) = 0. \tag{3.192}$$

We observe that, by our conditions, for any $c_1 > 0$ one can choose $c_2 > 0$, so that $f(u) \geq c_1 u - c_2$, for all $u > 0$. We choose c_1, so that $c_1\lambda > 2\mu_1$ for all $\lambda \in I$, and then select a corresponding c_2. Multiply the equation (3.172) by ϕ_1, and integrate:

$$\mu_1 \int_0^1 u\phi_1 \, dx \geq c_1\lambda \int_0^1 u\phi_1 \, dx - c_3 \geq 2\mu_1 \int_0^1 u\phi_1 \, dx - c_3,$$

with some constant $c_3 > 0$. It follows that

$$\int_0^1 u\phi_1 \, dx \leq \frac{c_3}{\mu_1}. \tag{3.193}$$

Assume now that we have a sequence of unbounded (in $C(0,1)$) solutions of (3.172), along some sequence of λ's in I. (Otherwise there is nothing to prove.) By above, $u''(0)$ would have to tend to infinity along this sequence of λ's. We show, next, that $u(x)$ must then tend to infinity on a subinterval of $(0,1)$, contradicting (3.193).

Observe that, by Lemma 3.21, $u(x)$ changes concavity exactly once between 0 and the point of global maximum, call it x_0, and $u(x)$ is strictly increasing on $(0, x_0)$. We may assume that $x_0 \geq \frac{1}{2}$, otherwise we can argue from the right end. Consider, say, $u(1/4)$. If $u(1/4)$ is large, then $u(x)$ is large on the interval $(1/4, 1/2)$, and then the integral $\int_0^1 u\phi_1 \, dx$ would get large, contradicting (3.193). So $u(1/4)$ cannot get large. We claim that $u''(1/4)$ cannot get large (positive) either. Indeed, $u''(x)$ is a convex function on $(0,1)$, and it is negative in the middle part of that interval. So, if $u''(1/4)$ was large, and with $u''(0)$ being large, $u''(x)$ would have to be large on $(0, 1/4)$, and then $u(x)$ would get large on say $(1/8, 1/4)$, resulting in the same contradiction with (3.193). But, if $u''(1/4)$ is bounded, by say M, we see (by convexity of u'') that $u''(x)$ is bounded by M on $(1/4, x_0)$, and we conclude that $u(x_0)$ is bounded, contradicting our assumption. $\Diamond$

3.11.2 *Exact multiplicity of solutions*

The following is our main exact multiplicity result. It shows that, for convex $f(u)$, the solution curve is similar to that for the corresponding second order equation.

Theorem 3.24. *Assume that $f(u) \subset C^2(0, \infty) \cap C[0, \infty)$ satisfies $f(u) > 0$ for $u \geq 0$, $\lim_{u \to \infty} \frac{f(u)}{u} = \infty$, $f'(0) \geq 0$, and $f''(u) > 0$ for $u > 0$. Then all positive solutions of (3.172) lie on a unique smooth curve of solutions. This curve starts at $(\lambda = 0, u = 0)$, it continues for $\lambda > 0$, until a critical λ_0, where it bends back, and continues for decreasing λ without any more turns, tending to infinity as $\lambda \downarrow 0$. In other words, we have exactly two, exactly one or no solution, depending on whether $0 < \lambda < \lambda_0$, $\lambda = \lambda_0$, or $\lambda > \lambda_0$. Moreover, all solutions are symmetric with respect to the midpoint $x = \frac{1}{2}$, and the maximum value of the solution, $u(\frac{1}{2})$, is strictly monotone on the curve.*

Proof: We begin with the trivial solution $(\lambda = 0, u = 0)$. At this point, the Implicit Function Theorem applies, giving us solutions for small $\lambda > 0$. These solutions are positive, since our problem is known to have a positive Green's function, see P. Korman [77]. We now continue the solution curve for increasing λ. We cannot continue this curve indefinitely. Indeed since $f(u)$ is positive, increasing and superlinear, we see that $f(u) \geq \alpha u$, for some $\alpha > 0$ and all $u > 0$. Multiplying the equation (3.172) by ϕ_1 and integrating, we get an upper bound for λ. Let, then, λ_0 be the supremum of λ's, for which we can continue the solution curve for increasing λ. Since, by Lemma 3.26, solutions are bounded for all $\lambda < \lambda_0$, passing to the limit in the integral form of our problem (3.172), we see that our problem has a solution u_0 at λ_0. So that the solution curve reaches a critical solution (λ_0, u_0), at which the linearized problem (3.176) has a non-trivial solution (since the Implicit Function Theorem is not applicable at (λ_0, u_0)).

We now recast our problem (3.172) in an operator form. Let $C_{00}^4(0, 1)$ denote the subspace of functions of $C^4(0, 1) \cap C[0, 1]$, that satisfy $u(0) = u'(0) = u(1) = u'(1) = 0$. Let $F(\lambda, u) : C_{00}^4(0, 1) \to C(0, 1)$ be defined as $F(\lambda, u) = u''''(x) - \lambda f(u(x))$. Then $F(\lambda_0, u_0) = 0$, and we shall apply the Crandall-Rabinowitz Theorem 1.7 to continue the solution curve locally around the critical point (λ_0, u_0). By Lemma 3.16, we see that the null space $N(F_u(\lambda_0, u_0)) = span\{w(x)\}$ is one dimensional, and then $codim\, R(F_u(\lambda_0, u_0)) = 1$, since $F_u(\lambda_0, u_0)$ is a Fredholm operator of index zero. To apply the Crandall-Rabinowitz Theorem 1.7, it remains to check that $F_\lambda(\lambda_0, u_0) \notin R(F_u(\lambda_0, u_0))$. Assuming the contrary, would imply existence of $v(x) \not\equiv 0$, such that

$$v''''(x) - \lambda_0 f_u(u_0(x))v = f(u_0(x)), \quad x \in (0, 1) \qquad (3.194)$$
$$v(0) = v'(0) = v(1) = v'(1) = 0.$$

From (3.194) and the linearized problem (3.176), we easily conclude that

$$\int_0^1 f(u_0(x))w(x)\, dx = 0.$$

But, by Lemma 3.19, $\int_0^1 f(u_0(x))w(x)\, dx = \frac{1}{4\lambda}u_0''(1)w''(1)$, and $u_0''(1) \neq 0$ by Lemma 3.17, while $w''(1) \neq 0$ by Lemma 3.16. This is a contradiction, and hence the Crandall-Rabinowitz Theorem 1.7 applies at (λ_0, u_0). The same analysis applies at any other singular solution. We conclude that all solution curves can be continued globally, i.e., at any point either the Implicit Function Theorem, or the Crandall-Rabinowitz Theorem 1.7 applies.

We show, next, that at (λ_0, u_0), and any other critical point, a turn to the left must occur. Near the point (λ_0, u_0), according to the Crandall-Rabinowitz Theorem 1.7, we have $\lambda = \lambda(s)$, $u = u(s)$, with $\lambda(0) = \lambda_0$, $\lambda'(0) = 0$, and $u(0) = u_0$, $u_s(0) = w$. Differentiating the equation (3.172) twice in s, and then setting $s = 0$, we have

$$u_{ss}''''(x) = \lambda_0 f'(u_0)u_{ss} + \lambda_0 f''(u_0)w^2 + \lambda''(0)f(u_0), \quad x \in (0,1)\ (3.195)$$

$$u_{ss}(0) = u_{ss}'(0) = u_{ss}(1) = u_{ss}'(1) = 0 \,.$$

Multiplying the equation in (3.195) by w, the equation in (3.176) by u_{ss}, subtracting the results, and integrating, we express

$$\lambda''(0) = -\frac{\int_0^1 f''(u_0)w^3\, dx}{\int_0^1 f(u_0)w\, dx} < 0\,,$$

since $f(u)$ is convex, and $w(x)$ is positive.

We now complete the proof. The curve of solutions, starting at $(\lambda = 0, u = 0)$, after turning back at (λ_0, u_0), continues without any more turns. Since, by Lemma 3.23, the maximum value of the solution, $u(1/2, \lambda)$, is strictly increasing on the curve, and the curve continues globally, we see that $u(1/2, \lambda)$ tends to infinity along the curve. By Lemma 3.26, this may happen only when $\lambda \downarrow 0$. Observe that $u(1/2, \lambda)$ changes from zero to infinity along this curve, and hence, by Lemma 3.23, this curve exhausts the set of all possible solutions (since it takes up all possible maximal values of solutions). $\diamond$

We also have the following uniqueness result, whose proof follows the familiar pattern, see P. Korman [91] for the proof. (The uniqueness part was previously proved in R. Dalmasso [42]. However, the bifurcation approach provides a simpler proof, and some extra information on the solution curve.)

Theorem 3.25. *Assume that $f(u) \subset C^1(0,\infty) \cap C[0,\infty)$ satisfies*

$$0 < f(u) < uf'(u) \quad for \ u > 0. \tag{3.196}$$

Then for any $\lambda > 0$, the problem (3.172) has at most one positive solution. Moreover, all positive solutions are non-singular (i.e., the corresponding linearized problem (3.176) has only the trivial solution), they lie on a single smooth solution curve, they are symmetric with respect to the midpoint $x = \frac{1}{2}$, and the maximum value of the solution, $u(\frac{1}{2})$, is strictly monotone on the curve.

3.12   Steady states of convective equations

We consider a class of convective reaction-diffusion equations (for $u = u(x,t)$)

$$\begin{aligned}
u_t &= u_{xx} + \epsilon(g(u))_x + f(u), & 0 < x < L, \ t > 0 \tag{3.197} \\
u(0,t) &= u(L,t) = 0, & t > 0 \\
u(x,0) &= u_0(x), & 0 < x < L.
\end{aligned}$$

Such equations occur in many physical applications, including models of gas dynamics in a tube, and non-Boussinesq convection, with nonlinear density dependence on the temperatures. If $g \equiv 0$, this is a well studied equation of reaction-diffusion type, while in case $f \equiv 0$, we have Burger's equation. An extensive study of the problem (3.197) appears in the papers of T.F. Chen, H.A. Levine and P.E. Sacks [34], and H.A. Levine, L.E. Payne, P.E. Sacks and B. Straughan [125]. In particular, they found that the dynamics of (3.197) can be substantially different from the reaction-diffusion case (when $g \equiv 0$). Most notably, the regions of blow up and stability need not be separated by a steady state.

The first step in studying long time behavior of solutions $u(x,t)$, involves understanding of the steady states, i.e., solutions of (here $u = u(x)$)

$$u'' + g(u)' + \lambda f(u) = 0 \ \text{on} \ (0,1), \ u(0) = u(1) = 0. \tag{3.198}$$

Here, for convenience, we take $L = 1$. We present a study of this problem, based on P. Korman [80]. We shall assume that $f(u) \in C^2(\bar{R}_+)$, λ is a positive parameter, and the function $g(u) \in C^2(\bar{R}_+)$ satisfies

$$g'(u) > 0, \ \text{for almost all} \ u > 0. \tag{3.199}$$

We shall follow the familiar approach. Before we prove positivity of solutions for the linearized problem, we need to understand the shape of positive solutions.

Lemma 3.27. *Any positive solution of (3.198) has only one point of local maximum (which is, of course, the point of global maximum).*

Proof: Assuming otherwise, the function $u(x)$ would have points of local minimum on $(0,1)$. Let x_1 be the largest such point, $u'(x_1) = 0$, and let $x_2 > x_1$ be such that $u(x_2) = u(x_1)$. Let $\bar{x}$ be the point of local maximum on (x_1, x_2), and denote $\bar{u} = u(\bar{x})$ and $u_1 = u(x_1)$. Multiply the equation (3.198) by u', and then integrate over $(x_1, \bar{x})$,

$$\int_{x_1}^{\bar{x}} g'(u)u'^2 \, dx + \lambda \int_{u_1}^{\bar{u}} f(u) \, du = 0. \tag{3.200}$$

Similarly, multiplying by u', and integrating over $(\bar{x}, x_2)$,

$$\frac{1}{2}u'^2(x_2) + \int_{\bar{x}}^{x_2} g'(u)u'^2 dx - \lambda \int_{u_1}^{\bar{u}} f(u) \, du = 0. \tag{3.201}$$

Adding (3.200) and (3.201), we obtain

$$\frac{1}{2}u'^2(x_2) + \int_{x_1}^{x_2} g'(u)u'^2 \, dx = 0,$$

which gives a contradiction, since the second term on the left is strictly positive. $\diamond$

Remark. Multiplying (3.198) by u', and integrating over $(0,1)$, one sees that $u'(0) > |u'(1)|$, and more generally, if $x_1 < x_2$, and $u(x_1) = u(x_2)$, then $u'(x_1) > |u'(x_2)|$. I.e., positive solutions are "skewed" to the left.

We shall denote by $x_0 \in (0, 1/2)$, the unique point of global maximum of $u(x)$. We shall need to consider the linearized problem for (3.198)

$$w'' + g'(u)w' + g''(u)u'w + \lambda f'(u)w = 0 \ \ \text{on} \ \ (0,1), \ \ w(0) = w(1) = 0, \tag{3.202}$$

and the adjoint of the linearized problem

$$\mu'' - g'(u)\mu' + \lambda f'(u)\mu = 0 \ \ \text{on} \ \ (0,1), \ \ \mu(0) = \mu(1) = 0. \tag{3.203}$$

Lemma 3.28. *If the problem (3.202) admits a nontrivial solution, it can be chosen so that $w(x) > 0$ on $(0,1)$.*

Proof: Differentiate the equation (3.198)

$$u_x'' + g'(u)u_x' + g''(u)u'u_x + \lambda f'(u)u_x = 0 \,. \qquad (3.204)$$

Let $p(x) = u''(x)w(x) - u'(x)w'(x)$. Multiplying the equation (3.204) by w, and subtracting from it the equation (3.202) multiplied by u_x, we express

$$p' + g'(u)p = 0, \quad \text{for } x \in (0,1) \,. \qquad (3.205)$$

Assume that $w(x)$ vanishes at some point in $(0,1)$. Assuming, first, that $w(x)$ has a root in $(0, x_0)$, we denote by $\xi \in (0, x_0)$ the smallest root of $w(x)$. We can choose $w(x) > 0$ on $(0, \xi)$, and then $p(0) = -u'(0)w'(0) < 0$, while $p(\xi) = -u'(\xi)w'(\xi) > 0$. But this is a contradiction, since $p(x)$ is a solution of the linear equation (3.205), and so it cannot change sign. Similarly, if $\xi \in (x_0, 1)$ is the largest root of $w(x)$, and $w(x) > 0$ on $(\xi, 1)$, then $p(\xi) > 0$ and $p(1) < 0$, which is again a contradiction. $\diamondsuit$

We also have the positivity property for the adjoint equation.

Lemma 3.29. *If the problem (3.202) admits a nontrivial solution, then so does the problem (3.203). Moreover, we can choose $\mu(x) > 0$ on $(0,1)$.*

Proof: We are given that zero is the principal eigenvalue of the operator $Lw = w'' + g'(u)w' + g''(u)u'w + \lambda f'(u)w$, with zero boundary conditions, and $w(x) > 0$ is the corresponding eigenfunction. If zero is not the principal eigenvalue of the transpose of L, then the principal eigenvalue is some $\theta \neq 0$, with the corresponding eigenfunction $z(x) > 0$, so that $(L'z = \theta z)$

$$z'' - g'(u)z' + \lambda f'(u)z = \theta z \quad \text{on } (0,1), \quad z(0) = z(1) = 0 \,. \qquad (3.206)$$

Multiplying (3.206) by w, and subtracting (3.202), multiplied by z, and then integrating over $(0,1)$, we obtain, after an integration by parts,

$$\int_0^1 w(x)z(x)\,dx = 0 \,,$$

a contradiction, since the integrand is positive. $\diamondsuit$

For the rest of this section, we shall consider a more special problem

$$u'' + au' + \lambda f(u) = 0 \quad \text{on } (0,1), \quad u(0) = u(1) = 0 \,, \qquad (3.207)$$

with a constant a, which we assume to be positive, without loss of generality (as seen by changing variables $x \to 1 - x$). We shall also need to consider the corresponding versions of (3.202) and (3.203):

$$w'' + aw' + \lambda f'(u)w = 0 \quad \text{on } (0,1), \quad w(0) = w(1) = 0 \,; \qquad (3.208)$$

$$\mu'' - a\mu' + \lambda f'(u)\mu = 0 \quad \text{on} \quad (0,1), \quad \mu(0) = \mu(1) = 0. \qquad (3.209)$$

We have the following exact multiplicity result.

Theorem 3.26. *Assume that* $f(u) \in C^2(\bar{R}_+)$ *satisfies the following conditions*

$$f''(u) > 0, \quad \text{for almost all } u > 0; \qquad (3.210)$$

$$f(u) \geq c_1 u^p + c_2, \quad \text{for all } u \geq 0, \qquad (3.211)$$

with constants $c_1, c_2 > 0$ *and* $p > 1$. *Then all positive solutions of (3.207) lie on a single smooth solution curve. This curve begins at* $(\lambda = 0, u = 0)$, *it continues for increasing* λ, *until a critical* $\lambda = \lambda_0 > 0$, *where it bends to the left, and continues without any more turns, tending to infinity as* $\lambda \to 0$. *Thus, the solution curve has two branches, referred to as lower (stable) branch* $u^-(x, \lambda)$, *and the upper (unstable) one* $u^+(x, \lambda)$. *We have, moreover,* $0 < u^-(x, \lambda) < u^+(x, \lambda)$ *for all* $0 < \lambda < \lambda_0$, *and* $x \in (0,1)$. *In particular, the problem (3.207) has exactly two, one or zero solutions, depending on whether* $0 < \lambda < \lambda_0$, $\lambda = \lambda_0$, *or* $\lambda > \lambda_0$.

Proof: The proof follows a familiar pattern, so we shall be sketchy, and more details can be found in P. Korman [80]. When $\lambda = 0$, there is a trivial solution $u = 0$. It follows by the Implicit Function Theorem that for $\lambda > 0$ and small, there is a smooth curve of solutions passing through $(\lambda = 0, u = 0)$. This solution curve cannot be continued indefinitely for increasing λ, since we can show that the problem (3.207) has no solutions for λ sufficiently large. By standard a priori estimates, the curve develops a singular solution at some (λ_0, u_0), see P. Korman [80] for details.

To show that the Crandall-Rabinowitz Theorem 1.7 applies at (λ_0, u_0), we rewrite the equation (3.207) in the operator form as

$$F(\lambda, u) \equiv u'' + au' + \lambda f(u) = 0,$$

where $F : R_+ \times C_0^2(0,1) \to C(0,1)$. Notice that $F_u(\lambda, u)w$ is given by the left hand side of (3.208). By the definition of λ_0, $F_u(\lambda_0, u_0)$ has to be singular, i.e., (3.208) has a nontrivial solution $w(x)$, which is positive by Lemma 3.28. The null-space $N(F_u(\lambda_0, u_0)) = \text{span}\{w(x)\}$ is one dimensional (it can be parameterized by $w'(0)$). It follows that $\text{codim}\, R(F_u(\lambda_0, u_0)) = 1$, since $F_u(\lambda_0, u_0)$ is a Fredholm operator of index zero. To apply the Crandall-Rabinowitz Theorem 1.7, it remains to check that $F_\lambda(\lambda_0, u_0) \notin R(F_u(\lambda_0, u_0))$. Assuming otherwise, would imply existence of $v(x) \not\equiv 0$, such that

$$v'' + av' + \lambda_0 f'(u_0)v = f(u_0), \quad 0 < x < 1, \quad v(0) = v(1) = 0. \qquad (3.212)$$

Multiplying (3.212) by μ, (3.209) by v, subtracting and integrating over $(0, 1)$, we obtain

$$\int_0^1 f(u_0)\mu \, dx = 0,$$

which is impossible, since both $f(u_0)$ and μ are positive.

Next, we compute the direction of bifurcation at (λ_0, u_0), and at any other turning point. Near the turning point (λ_0, u_0), we represent $\lambda = \lambda(s)$, $u = u(s)$, with $\lambda_0 = \lambda(0)$ and $u_0 = u(0)$. Differentiating the equation (3.207) in s twice, and setting $s = 0$, we obtain (using that $\lambda'(0) = 0, u_s(0) = w$)

$$u_{ss}'' + au_{ss}' + \lambda_0 f'(u)u_{ss} + \lambda_0 f''(u)w^2 + \lambda''(0)f(u) = 0,$$

$$u_{ss}(0) = u_{ss}(1) = 0.$$

Multiplying the last equation by μ, (3.209) by u_{ss}, integrating over $(0, 1)$ and subtracting, we have

$$\lambda''(0) = -\lambda_0 \frac{\int_0^1 f''(u)w^2\mu \, dx}{\int_0^1 f(u)\mu \, dx} < 0.$$

This implies that at (λ_0, u_0), as well as at any other turning point, the curve of solutions will bend to the left in (λ, u) "plane". Hence, there is only one turn on the solution curve. $\Diamond$

It turns out that almost all bounded solutions of the parabolic problem (3.197) converge to the lower (stable) steady state $u^-(x, \lambda)$, as $t \to \infty$, see P. Korman [80]. We mention that [80] has some results on multiplicity of positive steady state solutions for cubic $f(u)$, which involve considerable technical complications, since here solutions are not symmetric with respect to the mid-point of the interval.

3.13   Quasilinear boundary value problems

Following a recent paper of P. Korman and Y. Li [106], we study uniqueness and exact multiplicity of positive solutions for a class of quasilinear Dirichlet problems (here $u = u(x)$)

$$\varphi(u')' + \lambda f(u) = 0 \quad \text{for } -L < x < L, \quad u(-L) = u(L) = 0, \qquad (3.213)$$

depending on a positive parameter λ. The problem (3.213) has been studied extensively, but mostly under the assumptions consistent with the p-Laplacian case (i.e., $\varphi'(t) > 0$ for $t \neq 0$, while $\varphi(0) = \varphi'(0) = 0$). The equation is then degenerate elliptic. Instead, we shall assume that

$$\varphi \in C^2(R) \text{ is an odd function, and } \varphi'(t) > 0 \text{ for all } t \in R, \quad (3.214)$$

$$f(u) \in C^2(\bar{R}_+), \text{ and } f(u) > 0 \text{ for } u > 0. \quad (3.215)$$

Our main example will be $\varphi(t) = \frac{t}{\sqrt{1+t^2}}$, of the prescribed mean curvature equation, which was considered recently by P. Habets and P. Omari [67], D. Bonheure et al [22], and H. Pan [147]. We consider the problem on the interval $(-L, L)$ for convenience, which is related to the symmetry of solutions. By shifting, we can replace the interval $(-L, L)$ by any other interval (a, b) of length $2L$. Unlike the semilinear case, the equation (3.213) is not scaling invariant, which opens the possibility that bifurcation diagram may depend on the length of the interval. And, indeed the length of the interval matters, as was discovered in a recent paper of H. Pan [147]. For the prescribed mean curvature equation, and $f(u) = e^u$, H. Pan [147] showed that the bifurcation diagram is different for the cases $2L < \pi$, and $2L \geq \pi$. Using bifurcation theory analysis, P. Korman and Y. Li [106] had generalized the results of H. Pan [147] to any convex $f(u)$. The previous papers, mentioned above, were using mostly the time maps.

In addition to bifurcation diagram possibly depending on the length of the interval, there are two more new effects, i.e., properties different from semilinear equations. Here, a solution curve may develop an infinite slope at the end-points at some λ, and stop (which never happens for semilinear equations). In addition, solution curves may turn the "wrong way" (compared with semilinear equations). For example, if $f(u) = u^p$, $0 < p < 1$, and $\varphi(t) = \frac{t}{\sqrt{1+t^2}}$, of the prescribed mean curvature equation, P. Habets and P. Omari [68] have discovered that solution curve is parabola-like, with a turn to the left. This $f(u)$ is concave, so that for semilinear equations (i.e., when $\varphi(t) = t$) only turns to the right are possible. By contrast, we show that for convex $f(u)$, only turns to the left are possible for the prescribed mean curvature equation, similarly to semilinear equations.

To see what are the expected properties of the solution curves, it is convenient to work with an equivalent Dirichlet problem

$$u'' + \lambda h(u')f(u) = 0 \quad \text{for } -L < x < L, \ u(-L) = u(L) = 0, \quad (3.216)$$

where $h(t) = \frac{1}{\varphi'(t)}$. We shall assume that $f(u) \in C^2(\bar{R}_+)$, and

$$0 < h(t) \in C^2(R), \quad \text{and} \quad h(-t) = h(t) \quad \text{for all } t. \tag{3.217}$$

Without the term $h(u')$, we have a semilinear problem discussed previously. The extra term $h(u')$ may produce some new effects, especially in the case when the well-known Nagumo's condition is violated, i.e., when $h(u')$ is superquadratic in $|u'|$. Let us consider two examples.

Example Consider

$$u'' + \lambda \left(1 + u'^2\right) = 0 \quad \text{for } -1 < x < 1, \ u(-1) = u(1) = 0. \tag{3.218}$$

Integrating, we calculate the solution: $u(x, \lambda) = \frac{1}{\lambda} \ln \frac{\cos \lambda x}{\cos \lambda}$. Hence, for $0 < \lambda < \frac{\pi}{2}$, the problem (3.218) has a unique positive solution, while, as $\lambda \uparrow \frac{\pi}{2}$, the maximum value of solution, $u(0, \lambda)$, tends to infinity. For semilinear equations, similar behavior occurs if $f(u)$ is asymptotically linear (recall "bifurcation from infinity").

Example We consider a case when Nagumo's condition is violated

$$u'' + \lambda \left(1 + u'^4\right) = 0 \quad \text{for } -1 < x < 1, \ u(-1) = u(1) = 0. \tag{3.219}$$

We construct the solution of (3.219), by solving, first, the initial value problem

$$y'' + \lambda \left(1 + y'^4\right) = 0 \quad \text{for } -1 < x < 1, \ y(0) = y'(0) = 0, \tag{3.220}$$

and then selecting the constant $\alpha > 0$, so that $u(x) = y(x) + \alpha$ solves (3.219) (observe that $y(x) < 0$). Setting $y' = p(y)$, followed by $q = p^2$, we can integrate (3.220), obtaining

$$y' = -\sqrt{-\tan 2\lambda y}, \quad \text{for } x > 0. \tag{3.221}$$

One can probably finish the analysis of this problem in an elementary fashion, however it seems easier to rely on the results of the present section. We will show that all solutions of (3.219) lie on a unique solution curve, and the maximum value $\alpha = u(0, \lambda)$ is increasing, as we continue this curve, starting at $\lambda = 0$, $u = 0$, and that $\alpha = u(0, \lambda)$ increases without a bound along the curve. Since $u(1) = 0$, it follows that $y(1) = -\alpha$. As we continue the curve for increasing λ, the quantity $-2\lambda\alpha$ is monotone increasing, and hence at some λ_0, it will become equal to $\pi/2$. We then see from (3.221) that the solution $u_0 = u(x, \lambda)$ has infinite slope at both end points, while the maximum value of this solution is bounded. It follows from our results that the solution curve cannot be continued beyond the point (λ_0, u_0). This phenomenon of "sudden death" never happens for semilinear equations.

As we mentioned, (3.213) and (3.216) are two ways to write the same equation. When considering the form (3.216), we may allow for $h(t)$ to vanish at some points. The advantage of the form (3.213), is existence of first (or energy) integral. Indeed, defining $\Phi(z) = \int_0^z t\varphi'(t)\,dt$ and $F(z) = \int_0^z f(t)\,dt$, we see that any solution of (3.213) satisfies

$$\Phi(u'(x)) + \lambda F(u(x)) = constant. \tag{3.222}$$

Next, we discuss some properties of positive solutions. These properties are similar to those of semilinear equations, although proofs tend to be more involved. The following lemma on symmetry of solutions is included in the results of B. Gidas, W.-M. Ni and L. Nirenberg [63], but we provide a simple proof.

Lemma 3.30. *Assume that $h(t)$ satisfies (3.217). Then any positive solution of the problem (3.216) is an even function, with $u'(x) < 0$ for $x > 0$.*

Proof: It suffices to show that $u(x)$ is symmetric with respect to any of its critical points. Let $u'(\xi) = 0$ at some ξ. Consider $v(x) \equiv u(2\xi - x)$. We see that $v(x)$ is another solution of (3.216), with $v(\xi) = u(\xi)$, and $v'(\xi) = u'(\xi) = 0$. It follows that $u(x) \equiv v(x)$. This is only possible if there is a single critical point, $\xi = 0$. ◇

This lemma implies symmetry of positive solutions for the problem (3.213), as well.

The concept of energy is useful in proving the following lemma, see [106].

Lemma 3.31. *Under the conditions (3.214) and (3.215), any two positive solutions of (3.213) cannot intersect, and hence they are strictly ordered on $(-L, L)$.*

By Lemma 3.30, $\alpha \equiv u(0)$ is the maximum value of solution. As in the semilinear case, it is impossible for any two positive solutions of (3.213) to share the same maximum value α, although the proof is more involved here, since the scaling argument does not apply, see [106].

Lemma 3.32. *Assume that the conditions (3.214) and (3.215) hold. Then, the value of $u(0) = \alpha$ uniquely identifies the solution pair $(\lambda, u(x))$ of (3.213) (i.e., there is at most one λ, with at most one positive solution $u(x)$, so that $u(0) = \alpha$).*

To prove positivity for the linearized problem, it is more convenient to work with (3.216), for which the linearized problem takes the form

$$w'' + \lambda h'(u')f(u(x))w' + \lambda h(u')f'(u(x))w = 0, \text{ for } -L < x < L, \quad (3.223)$$
$$w(-L) = w(L) = 0.$$

Clearly, $w'(L) \neq 0$ for any non-trivial solution, and hence we may assume that

$$w'(L) < 0. \tag{3.224}$$

Lemma 3.33. *If the problem (3.223) admits a nontrivial solution, then it does not change sign, i.e., we may assume that $w(x) > 0$ on $(-L, L)$.*

Proof: Assume that $w(x)$ vanishes on $(0, L)$, and the other case when $w(x)$ vanishes on $(-L, 0)$ is similar. We can then find a $\xi \in (0, L)$, so that $w(\xi) = 0$, and $w(x) > 0$ on (ξ, L). Differentiate the equation in (3.216)

$$u_x'' + \lambda h'(u')f(u)u_x' + \lambda h(u')f'(u)u_x = 0. \tag{3.225}$$

Combining (3.225) with the equation in (3.223), we have

$$(u'w' - wu'')' + \lambda h'(u')f(u)(u'w' - wu'') = 0. \tag{3.226}$$

We see that the function $p(x) \equiv u'w' - wu''$ satisfies a linear first order equation. Hence, it is either of one sign, or identically zero. We have $p(L) = u'(L)w'(L) \geq 0$, in view of (3.224). On the other hand, $p(\xi) = u'(\xi)w'(\xi) < 0$, a contradiction. $\diamond$

We have the following general description of the solution curve, for any $L > 0$.

Theorem 3.27. *([106]) Assume that $\varphi(t)$ satisfies the condition (3.214), and the function $f(u) \in C^2(\bar{R}_+)$ is positive on $(0, \infty)$. Then all positive solutions of the problem (3.213), satisfying $|u'(L)| < \infty$, lie on a unique curve in the $(\lambda, u(0))$ plane. This curve is parameterized by $\alpha = u(0)$. If, moreover, we assume that*

$$f(u) \text{ is increasing for all } u > 0, \tag{3.227}$$
$$\text{and } \Phi(t) \text{ is uniformly bounded on } (-\infty, \infty),$$

and

$$\lim_{u \to \infty} \frac{f(u)}{F(u)} = 0, \tag{3.228}$$

then this curve cannot extend globally, i.e., it stops at some $(\bar{\lambda}, \bar{\alpha})$.

The linearized problem, corresponding to (3.213), is

$$(\varphi'(u')w')' + \lambda f'(u)w = 0, \quad \text{for } -L < x < L, \tag{3.229}$$
$$w(-L) = w(L) = 0.$$

Lemma 3.33 implies that any non-trivial solution of (3.229) may be assumed to be positive on $(-L, L)$. We have the following exact multiplicity result.

Theorem 3.28. *Assume that $\varphi(t)$ satisfies the condition (3.214), and*

$$t\varphi''(t) \leq 0, \quad \text{for all } t \in R. \tag{3.230}$$

Moreover, assume that the range of $\varphi(t)$ over R is bounded, while the function $f(u) \in C^2(\bar{R}_+)$ is convex, it satisfies $f'(u) > 0$ for $u > 0$, and it is bounded below by a positive constant on $[0, \infty)$. Then the problem (3.213) has at most two positive solutions for any $\lambda > 0$. Moreover, all positive solutions lie on a unique solution curve in the $(\lambda, u(0))$ plane. This curve begins at the point $(\lambda = 0, u(0) = 0)$, and either it tends to infinity at some $\lambda_0 > 0$, or at λ_0 it develops infinite slope at $x = \pm L$ and stops, or else it bends back at some $\lambda_0 > 0$. After the turn, the curve continues without any more turns, and it either tends to infinity for decreasing λ, or else it develops infinite slope at $x = \pm L$ and stops at some $\bar{\lambda}$, $0 \leq \bar{\lambda} < \lambda_0$ (the last possibility holds, provided the conditions (3.227) and (3.228) are satisfied).

Proof: The proof follows the familiar pattern, so we just show how to compute the direction of turn at any critical point $(\lambda_0, u_0(x))$. According to the Crandall-Rabinowitz Theorem 1.7, the solution set near the point $(\lambda_0, u_0(x))$ is a curve $\lambda = \lambda(s)$, $u = u(s)$, with $\lambda(0) = \lambda_0$ and $u(0) = u_0(x)$, and moreover,

$$\lambda'(0) = 0, \quad \text{and } u_s(0) = w(x), \tag{3.231}$$

where $w(x)$ is the solution of the linearized problem. To prove that only turns to the left are possible, we need to show that $\lambda''(0) < 0$. To obtain an expression for $\lambda''(s)$, we differentiate the equation (3.213) in s twice, obtaining

$$\left(\varphi'(u')u'_{ss} + \varphi''(u')u'^2_s\right)' + \lambda f'(u)u_{ss} + 2\lambda'f'(u)u_s + \lambda f''(u)u^2_s + \lambda''f(u) = 0.$$

Setting here $s = 0$, and using (3.231), we have (we write u instead of u_0)

$$(\varphi'(u')u'_{ss})' + \left(\varphi''(u')w'^2\right)' + \lambda_0 f'(u)u_{ss} + \lambda_0 f''(u)w^2 + \lambda''(0)f(u) = 0$$
$$u_{ss}(-L) = u_{ss}(L) = 0.$$

Multiplying this equation by w, the linearized equation (3.229) by u_{ss}, subtracting and then integrating, we express

$$\lambda''(0) = \frac{-\lambda_0 \int_{-L}^{L} f''(u)w^3 \, dx + \int_{-L}^{L} \varphi''(u')w'^3 \, dx}{\int_{-L}^{L} f(u)w \, dx}. \tag{3.232}$$

Next we observe that $w'(x) \leq 0$ on $[0, L]$. Indeed, assuming otherwise, we would get a contradiction in (3.213) at any point of local minimum of $w(x)$ (it is here that we use the assumption $f'(u) > 0$). On the interval $(0, L)$, $w'(x) \leq 0$, $u'(x) < 0$, and hence $\varphi''(u') \geq 0$, by (3.230). On $(-L, 0)$, these inequalities are all reversed. It follows that $\int_{-L}^{L} \varphi''(u')w'^3 \, dx \leq 0$, and we conclude that $\lambda''(0) < 0$. Hence, there is at most one turn (to the left) on any solutions curve. $\diamond$

We also have the following uniqueness result. We state it for the prescribed mean curvature equation, and $f(u) = u^q$, although clearly both $\varphi(t)$ and $f(u)$ can be more general. For $L = 1$ this result was established earlier in [22] and [67], while H. Pan [147] has observed that the same result holds for any L. The present statement, with some extra information on the solution curve, is from P. Korman and Y. Li [106].

Theorem 3.29. *([106]) Consider the problem*

$$\varphi(u')' + \lambda u^q = 0 \quad for \ -L < x < L, \ \ u(-L) = u(L) = 0\,, \tag{3.233}$$

where $\varphi(t) = \frac{t}{\sqrt{1+t^2}}$, with any constants $q > 1$ and $L > 0$, and λ is a positive parameter. Then all positive solutions lie an a unique curve in $(\lambda, \alpha = u(0))$ plane. This curve begins at some $(\bar{\lambda}, \bar{\alpha})$, and tends to $(\lambda = \infty, \alpha = 0)$, without any turns. I.e., for $\lambda \in (0, \bar{\lambda})$ the problem has no positive solutions, and for $\lambda \geq \bar{\lambda}$ it has exactly one positive solution. The solution at $\lambda = \bar{\lambda}$ has an infinite slope at $x = \pm L$.

We consider, next, the prescribed mean curvature equation with exponential nonlinearity

$$\varphi(u')' + \lambda e^u = 0 \quad for \ -L < x < L, \ \ u(-L) = u(L) = 0\,, \tag{3.234}$$

where $\varphi(t) = \frac{t}{\sqrt{1+t^2}}$. We recall the following nice result of H. Pan [147] (which was given a different proof, and generalized in [106]).

Theorem 3.30. *([147]) All solutions of the problem (3.234) lie on a unique solution curve. This curve begins at $(\lambda = 0, u = 0)$ and continues for $\lambda > 0$. It makes exactly one turn at some $\lambda = \lambda_0$. If $2L \geq \pi$, then after the turn the curve continues for all $\lambda > 0$, and $u(0) \to \infty$ as $\lambda \to 0$. In case $2L < \pi$, after the turn, we have a solution with $|u'(\pm L)| = \infty$ at some $\bar{\lambda} < \lambda_0$, and the solution curve stops at $\bar{\lambda}$.*

3.13.1 Numerical computations for the prescribed mean curvature equation

Following P. Korman and Y. Li [106], we discuss numerical computation of positive solutions for the problem

$$\varphi(u')' + \lambda f(u) = 0 \quad \text{for } -L < x < L, \ u(-L) = u(L) = 0, \qquad (3.235)$$

where $\varphi(t) = \frac{t}{\sqrt{1+t^2}}$, and $f(u) > 0$ for $u > 0$. As we saw above, the maximum value $\alpha = u(0)$ uniquely identifies the solution pair $(\lambda, u(x))$. Since solutions of (3.235) are even, and $u'(0) = 0$, the knowledge of λ and $\alpha = u(0)$ allows a quick and easy computation of solution $u(x)$, by putting the problem (3.235) into an equivalent form (3.216), with $h(t) = (1 + t^2)^{3/2}$, and solving the initial value problem, starting at $x = 0$ (in *Mathematica*, this requires just one command **NDSolve**). It follows that a two-dimensional curve $(\lambda, u(0))$, the *solution curve*, gives a faithful representation of the set of all positive solutions for the problem (3.235), for all values of λ. The main question is: given $\alpha = u(0)$, how does one compute the corresponding λ? (Recall that for semilinear equations, this is done by "shooting" and scaling, and we cannot do scaling here.)

We begin computation of the solution curve, with the energy relation (3.222), which for the prescribed mean curvature equation takes the form (here $\Phi(t) = 1 - \frac{1}{\sqrt{1+t^2}}$)

$$\frac{1}{\sqrt{1 + u'^2}} = 1 - \lambda \Delta F, \qquad (3.236)$$

where we denote $F(u) = \int_0^u f(t)\,dt$, and $\Delta F = F(\alpha) - F(u)$. Solving for $u'(x)$, with $x \in (0, L)$, we get

$$\frac{du}{dx} = -\frac{\sqrt{2\lambda \Delta F - \lambda^2 (\Delta F)^2}}{1 - \lambda \Delta F}.$$

Separating variables, and integrating over $x \in (0, L)$,

$$\int_0^\alpha \frac{1 - \lambda \Delta F}{\sqrt{2\lambda \Delta F - \lambda^2 (\Delta F)^2}}\,du = L. \qquad (3.237)$$

This is a relation, connecting λ and α, i.e., given α we can compute the corresponding unique λ. We scale $u = \alpha v$, and put (3.237) into the form

$$\alpha g(\alpha, \lambda) - L = 0, \qquad (3.238)$$

where $g(\alpha, \lambda) \equiv \int_0^1 \frac{1-\lambda\Delta F}{\sqrt{2\lambda\Delta F - \lambda^2(\Delta F)^2}}\, dv$, with a redefined $\Delta F = F(\alpha) - F(\alpha v)$. We now vary α's, with a certain step-size, and compute the corresponding λ from (3.238), by employing the Newton's method

$$\alpha\left[g(\alpha, \lambda_n) + g_\lambda(\alpha, \lambda_n)(\lambda_{n+1} - \lambda_n)\right] = L,$$

i.e.,

$$\lambda_{n+1} = \lambda_n + \frac{L/\alpha - g(\alpha, \lambda_n)}{g_\lambda(\alpha, \lambda_n)}.$$

For the initial guess λ_0, we took the value of λ, which was computed for the preceding α. When $\lambda F(\alpha) = 1$, we see from the energy relation (3.236) that solution has infinite derivatives at $x = \pm L$. We, therefore, stop calculating the solution curve, once the inequality $\lambda F(\alpha) < 1$ is violated.

We now describe numerical computations in case of exponential $f(u)$, i.e., for the problem (3.234). According to the energy relation, solutions with infinite slope at the end-points occur when $\lambda F(\alpha) = 1$. This is a curve in (λ, α) plane, which we call *the blow up curve*. For $f(u) = e^u$, the blow up curve is

$$\alpha = \ln\left(1 + \frac{1}{\lambda}\right). \tag{3.239}$$

Example 1 Using *Mathematica*, we have solved the problem (3.234), for $L = 1$. The solution curve touches the blow up curve (3.239) (dashed), and stops, see Figure 3.8.

Example 2 We have solved the problem (3.234), for $L = 2$. The solution curve stays below the blow up curve (3.239) (dashed), and it continues for all λ, as $\lambda \to 0$, see Figure 3.9.

3.14 The time map for quasilinear equations

Following T. Adamowicz and P. Korman [1], we now present a generalization of the time map formula of R. Schaaf [161], to quasilinear equations, including the case of p-Laplacian.

We wish to solve the initial value problem

$$\varphi(u'(x))' + f(u(x)) = 0 \quad u(0) = 0, \ u'(0) = r > 0, \tag{3.240}$$

and calculate the first root T. The map $T = T(r)$ is commonly referred to as the *time map*. We assume that the function $\varphi(t) \in C^1(R)$ is odd, and it satisfies

$$\varphi'(t) > 0 \quad \text{for all } t \in R. \tag{3.241}$$

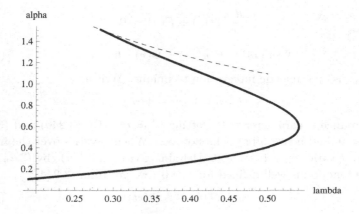

Fig. 3.8   Solution curve of the problem (3.234), when $L = 1$

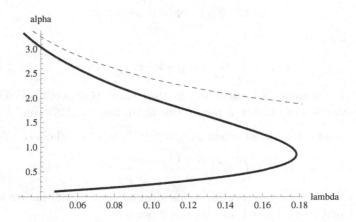

Fig. 3.9   Solution curve of the problem (3.234), when $L = 2$

We also assume that $f(u) \in C(R)$ satisfies
$$f(u) > 0 \quad \text{for all } u \geq 0. \tag{3.242}$$
It is easy to see that the solution $u(x)$ is symmetric with respect to the point of maximum. So that at $x = T/2$, the solution achieves its maximum, which we denote by α, $\alpha = u(T/2)$. We denote $\Psi(u) = \int_0^u t\varphi'(t)\,dt$, $\Psi'(u) = u\varphi'(u)$, and as usual $F(u) = \int_0^u f(t)\,dt$. We multiply the equation (3.240) by u', and transform it as follows:
$$u'(x)\varphi'(u'(x))\,u''(x) + f(u(x))u'(x) = 0\,;$$

$$\frac{d}{dx}\left[\Psi(u') + F(u)\right] = 0\,;$$

$$\Psi(u'(x)) + F(u(x)) = \Psi(r), \quad \text{for all } x\,. \tag{3.243}$$

We now use parametric integration technique. Writing

$$F(u) = \Psi(r) - \Psi(u')\,, \tag{3.244}$$

we introduce a parameter t, by letting $u' = rt$. (By (3.240) and (3.241), $u''(x) < 0$, and hence $u'(x)$ is monotone.) When x varies over the interval $[0, T/2]$, t varies over $1 \geq t \geq 0$. Defining $G(u) = F^{-1}(u)$ (by (3.242), the inverse function is well defined for $u \geq 0$), we get from (3.244)

$$u = G\left(\Psi(r) - \Psi(rt)\right)\,. \tag{3.245}$$

Then

$$dx = \frac{du}{u'(x)} = \frac{G'(\Psi(r)-\Psi(rt))\left(-rt\varphi'(rt)r\right)}{rt}\,dt \tag{3.246}$$
$$= -G'\left(\Psi(r) - \Psi(rt)\right)\varphi'(rt)r\,dt\,.$$

Integrating,

$$T/2 = \int_0^1 G'\left(\Psi(r) - \Psi(rt)\right)\varphi'(rt)r\,dt\,. \tag{3.247}$$

We have obtained a formula for the time map. Moreover, one can easily get the solution of (3.240) in parametric form, from (3.245) and (3.246).

In case of p-Laplacian, when $\varphi(t) = t|t|^{p-2}$, the formula becomes

$$T/2 = (p-1)r^{p-1}\int_0^1 G'\left(\frac{p-1}{p}r^p - \frac{p-1}{p}r^p t^p\right)t^{p-2}\,dt\,. \tag{3.248}$$

When $p = 2$, it is easy to transform this formula into the well-known formula of R. Schaaf [161]. Indeed, in case $p = 2$, we have

$$T = 2\int_0^1 G'\left(\tfrac{1}{2}r^2 - \tfrac{1}{2}r^2 t^2\right)r\,dt \tag{3.249}$$
$$= 2\int_0^{\pi/2} G'\left(\tfrac{1}{2}r^2\sin^2\theta\right)r\sin\theta\,d\theta\,,$$

after a change of variables, $t = \cos\theta$. R. Schaaf [161] defines a function $g(x)$ by the relation

$$g(x) = F^{-1}\left(\frac{1}{2}x^2\right) = G\left(\frac{1}{2}x^2\right)\,.$$

Clearly,

$$g'(x) = G'\left(\frac{1}{2}x^2\right)x\,,$$

and so (3.249) gives us

$$T = 2 \int_0^{\pi/2} g'(r \sin \theta) \, d\theta = \int_0^{\pi} g'(r \sin \theta) \, d\theta \,,$$

which is R. Schaaf's time map formula, see [161].

Let us now solve a parameter dependent problem for the p-Laplacian, with $\varphi(t) = t|t|^{p-2}$,

$$\varphi(u'(x))' + \lambda f(u(x)) = 0, \quad \text{for } x \in (-1,1), \quad u(-1) = u(1) = 0 \,. \quad (3.250)$$

Let $\alpha = u(0)$ denote the maximum value of solution. We wish to draw the bifurcation diagram for this problem in (λ, α) plane. Consider an initial value problem

$$\varphi(v'(\xi))' + f(v(\xi)) = 0, \quad v(0) = \alpha, \quad v'(0) = 0 \,. \quad (3.251)$$

The change of variables, $x = \frac{1}{\lambda^{1/p}}\xi$, transforms the equation in (3.250) into the one in (3.251). If R is the first root of v, then $R = \lambda^{1/p}$, i.e.,

$$\lambda = R^p = (T/2)^p \,, \quad (3.252)$$

since $2R$ gives the time map. From the energy relation (3.243), $F(\alpha) = \Psi(r)$, i.e.,

$$r = \Psi^{-1}(F(\alpha)) \,. \quad (3.253)$$

Starting with an α, we compute r by (3.253), then T by (3.248), and finally λ from (3.252). Numerically, for a mesh of α_i's, we compute this way corresponding λ_i's, and then plot the points (λ_i, α_i), to get the bifurcation diagram.

Example Consider problem (3.250), in the case of p-Laplacian, and $f(u) = e^u$. Compute $F(u) = e^u - 1$, $G(u) = \ln(u+1)$. The formula (3.253) becomes

$$r = \left[\frac{p}{p-1} (e^\alpha - 1) \right]^{1/p} \,.$$

The formula (3.248) takes the form

$$T/2 = (p-1)r^{p-1} \int_0^1 \frac{t^{p-2}}{\frac{p-1}{p}r^p - \frac{p-1}{p}r^p t^p + 1} \, dt \,.$$

Mathematica is able to compute this integral through hypergeometric functions, for any $p > 1$. Combining these formulas with (3.252), we compute the bifurcation diagram. In Figure 3.10, we present the computed bifurcation diagram for the case $p = \frac{3}{2}$. It is similar to the one for the semilinear case, $p = 2$.

We remark that in the same paper, T. Adamowicz and P. Korman [1] have developed yet another generalization of R. Schaaf's time map formula.

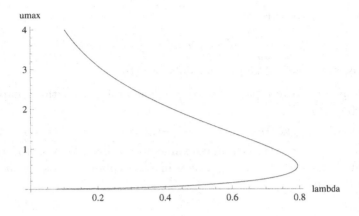

Fig. 3.10   Bifurcation diagram for $\frac{3}{2}$-Laplacian, and $f(u) = e^u$

3.15   Uniqueness for a p-Laplace case

Following P. Korman [101], we now give an exact description of the curve of positive solutions for the p-Laplace problem in the one-dimensional case

$$\varphi(u'(x))' + \lambda f(u(x)) = 0 \quad \text{for } -1 < x < 1, \ u(-1) = u(1) = 0, \quad (3.254)$$

where $\varphi(t) = t|t|^{p-2}$, $p > 1$, λ a positive parameter, and the function $f(u) \in C^1(\bar{R}_+)$ satisfies

$$f(u) < 0, \ \text{for } u \in (0, \gamma), \ f(u) > 0, \ \text{for } u > \gamma, \quad (3.255)$$

for some $\gamma > 0$, and

$$f'(u) - (p-1)\frac{f(u)}{u} > 0, \ \text{for } u > \gamma. \quad (3.256)$$

This equation is autonomous, and scaling invariant, and so the problem can be posed on any interval. Here we chose the interval $(-1, 1)$ for convenience, related to the symmetry of solutions. In case $p = 2$, we have considered this class of equations before. There are some complications in case $p \neq 2$.

The following lemma gives a known condition for the Hopf's lemma to hold, see J.L. Vázquez [185]. Moreover, in the present ODE case a completely elementary proof, using the Gronwall's lemma, can be easily given.

Lemma 3.34. *Assume that* $f(u) \in C^1(\bar{R}_+)$ *satisfies* $f(0) = 0$, *and*

$$-f(s) < c_0 s^{p-1}, \ \text{for small } s > 0, \ \text{and some } c_0 > 0. \quad (3.257)$$

If $u(x)$ is a positive solution of (3.254), then

$$u'(1) < 0.$$

We shall need the linearized problem for (3.254)

$$(\varphi'(u'(x))w'(x))' + \lambda f'(u(x))w(x) = 0 \quad \text{for } -1 < x < 1, \quad (3.258)$$
$$w(-1) = w(1) = 0.$$

Recall that we call a solution $u(x)$ of (3.254) *non-singular*, if the corresponding problem (3.258) admits only the trivial solution $w = 0$, otherwise $u(x)$ is a *singular* solution. We have the following precise description of the solution curve. (The uniqueness part was proved previously by J. Cheng [35].)

Theorem 3.31. *Assume that the conditions (3.255) and (3.256) hold, and $\lim_{u\to\infty} \frac{f(u)}{u^q} > 0$ for some $q > \min(1, p-1)$. In case $f(0) = 0$, we assume additionally that (3.257) holds. Then the problem (3.254) has at most one positive solution for any λ, while for λ sufficiently small, the problem (3.254) has a unique positive solution. Moreover, all positive solutions of (3.254) are non-singular, they lie on a unique solution curve, extending for $0 < \lambda \le \lambda_0 \le \infty$.*

In case $f(0) < 0$, we have $\lambda_0 < \infty$, there are no turns on the solution curve, and we have $u(0, \lambda) \to \infty$ as $\lambda \to 0$, and $u'(\pm 1, \lambda_0) = 0$. I.e., for $\lambda > \lambda_0$ there are no positive solutions, while for $0 < \lambda \le \lambda_0$, there is a unique positive non-singular solution.

In case $f(0) = 0$, we have $\lambda_0 = \infty$, there are no turns on the solution curve, with $u(0, \lambda) \to \infty$ as $\lambda \to 0$, and $\lim_{\lambda\to\infty} u(x, \lambda) = 0$, for $x \ne 0$. I.e., for all $\lambda > 0$, there is a unique positive non-singular solution.

What is remarkable here is that essentially no restrictions are placed on $f(u)$ in the region where it is negative. We shall present the proof, after a series of lemmas, most of which hold under considerably more general conditions.

Lemma 3.35. *Assume that $f(u) \in C^1(\bar{R}_+)$. Then any positive solution of (3.254) is an even function, with $u'(x) < 0$ for $x > 0$.*

Proof: If $\xi > 0$ is a critical point of $u(x)$, i.e., $u'(\xi) = 0$, then the solutions $u(x)$ and $u(2\xi - x)$ have the same Cauchy data, and so by uniqueness for initial value problems, the graph of $u(x)$ is symmetric with respect to ξ, which is impossible, in view of the boundary conditions. It follows that $x = 0$ is the only critical point, the point of global maximum. $\diamond$

Lemma 3.36. *Assume that $f(u) \in C^1(\bar{R}_+)$, while $u(x)$ and $w(x)$ are any solutions of (3.254) and (3.258) respectively. Then*

$$\varphi'(u')\,(u''w - u'w') = constant, \quad for \; all \; x \in [-1, 1]. \tag{3.259}$$

Proof: Just differentiate the left hand side of (3.259), and use the equations (3.254) and (3.258). $\diamond$

Lemma 3.37. *Assume that $f(u) \in C^1(\bar{R}_+)$. If the linearized problem (3.258) admits a non-trivial solution, then it does not change sign on $(-1, 1)$, i.e., we may assume that $w(x) > 0$ on $(-1, 1)$.*

Proof: Assume on the contrary that $w(x)$ vanishes on $(0, 1)$ (the case when $w(x)$ vanishes on $(-1, 0)$ is similar). Let $\xi > 0$ be the largest root of $w(x)$ on $(0, 1)$. We may assume that $w(x) > 0$ on $(\xi, 1)$, and then $w'(\xi) > 0$. Evaluating the expression in (3.259) at $x = \xi$, and at $x = 1$,

$$\varphi'(u'(\xi))u'(\xi)w'(\xi) = \varphi'(u'(1))u'(1)w'(1).$$

The quantity on the left is negative, while the one on the right is non-negative, a contradiction. $\diamond$

Given $u(x)$ and $w(x)$, solutions of (3.254) and (3.258) respectively, we again consider the following function, motivated by M. Tang [182]

$$T(x) = x\left[(p - 1)\varphi(u'(x))w'(x) + \lambda f(u(x))w(x)\right] - (p - 1)\varphi(u'(x))w(x).$$

The following lemma is proved by a direct computation.

Lemma 3.38. *Assume that $f(u) \in C^1(\bar{R}_+)$, while $u(x)$ and $w(x)$ are any solutions of (3.254) and (3.258) respectively. Then*

$$T'(x) = p\lambda f(u(x))w(x). \tag{3.260}$$

By Lemma 3.35, $u(0)$ gives the maximum value of the solution. If we now assume that $f(u)$ satisfies the condition (3.255), then it is easy to see that $f(u(0)) > 0$ (otherwise we get a contradiction, multiplying (3.254) by u, and integrating over $(-1, 1)$). So that we can find a point $x_0 \in (0, 1)$, such that

$$u(x_0) = \gamma.$$

(I.e., $f(u(x)) > 0$ on $(0, x_0)$, and $f(u(x)) < 0$ on $(x_0, 1)$.) Define

$$q(x) = (p-1)(1-x)\varphi(u'(x)) + \varphi'(u'(x))u(x).$$

Lemma 3.39. *Assume that* $f(u) \in C^1(\bar{R}_+)$ *satisfies the condition (3.255). Then*

$$q(x_0) < 0.$$

Proof: We have $(p-1)\varphi(t) = t\varphi'(t)$, and so we can rewrite

$$q(x) = \varphi'(u'(x))\left[(1-x)u'(x) + u(x)\right].$$

Since $\varphi'(t) > 0$ for all $t \neq 0$, it suffices to show that $r(x) \equiv (1-x)u'(x) + u(x) < 0$ on $[x_0, 1)$. We have $r(1) = 0$, and

$$r'(x) = (1-x)u''(x) = -(1-x)\lambda\frac{f(u(x))}{\varphi'(u'(x))} > 0, \quad \text{on } (x_0, 1),$$

and the proof follows. $\diamond$

Lemma 3.40. *Assume that* $f(u) \in C^1(\bar{R}_+)$ *satisfies the conditions (3.255) and (3.256), while* $u(x)$ *and* $w(x)$ *are any solutions of (3.254) and (3.258) respectively. Then*

$$(p-1)w(x_0)\varphi(u'(x_0)) - u(x_0)w'(x_0)\varphi'(u'(x_0)) > 0, \qquad (3.261)$$

which implies, in particular, that

$$w'(x_0) < 0. \qquad (3.262)$$

Proof: By a direct computation

$$[(p-1)w(x)\varphi(u'(x)) - u(x)w'(x)\varphi'(u'(x))]' = \lambda\left[f'(u) - (p-1)\frac{f(u)}{u}\right]uw.$$

The quantity on the right is positive on $(0, x_0)$, in view of our conditions, and Lemma 3.37. Integrating over $(0, x_0)$, we conclude (3.261). $\diamond$

We have all the pieces in place for the following crucial lemma.

Lemma 3.41. *Under the conditions (3.255) and (3.256), any positive solution of (3.254) is non-singular, i.e., the corresponding linearized problem (3.258) admits only the trivial solution.*

Proof: Assuming the contrary, let $w(x) > 0$ on $(-1, 1)$ be a solution of (3.258) (positivity follows by Lemma 3.37). By Lemma 3.36, and since $f(u(x_0)) = 0$, we have

$$\varphi'(u'(1))u'(1)w'(1) = \varphi'(u'(x_0))u'(x_0)w'(x_0) \qquad (3.263)$$
$$= (p-1)\varphi(u'(x_0))w'(x_0).$$

Integrating (3.260) over $(x_0, 1)$, we have

$$T(1) - T(x_0) = p\lambda \int_{x_0}^{1} f(u(x))w(x)\, dx < 0,$$

i.e.,

$$L \equiv (p-1)\varphi(u'(1))w'(1) - (p-1)x_0\varphi(u'(x_0))w'(x_0)$$
$$+ (p-1)\varphi(u'(x_0))w(x_0) < 0.$$

On the other hand, using (3.263), then (3.261), followed by (3.262) and Lemma 3.39, we have

$$L = (p-1)\varphi(u'(x_0))w'(x_0) - (p-1)x_0\varphi(u'(x_0))w'(x_0)$$
$$+ (p-1)\varphi(u'(x_0))w(x_0) > (p-1)\varphi(u'(x_0))w'(x_0)$$
$$- (p-1)x_0\varphi(u'(x_0))w'(x_0) + u(x_0)w'(x_0)\varphi'(u'(x_0)) = w'(x_0)q(x_0) > 0,$$

a contradiction. $\diamondsuit$

This lemma implies that there are no turns on the solution curves. For a more detailed description of the bifurcation diagram, we need to establish some additional properties of solutions. Denoting $F(u) = \int_0^u f(t)\, dt$, one shows by differentiation that

$$E(x) \equiv \frac{p-1}{p}|u'(x)|^p + \lambda F(u(x)) = constant, \quad \text{for all } x \in [-1, 1]. \quad (3.264)$$

It is customary to think of $E(x)$ as "energy".

The following lemma is variation on what we had before, in case $p = 2$.

Lemma 3.42. *The maximum value of the positive solution of (3.254), $u(0) = \alpha$, uniquely identifies the solution pair $(\lambda, u(x))$ (i.e., there is at most one λ, with at most one solution $u(x)$, so that $u(0) = \alpha$).*

Proof of the Theorem 3.31 We begin by showing that the problem (3.254) has solutions for some $\lambda > 0$. Observe that our condition (3.256) implies that the function $\frac{f(u)}{u^{p-1}}$ is increasing, i.e., $f(u) \to \infty$ for large u.

Let $\theta > \gamma$ be the point where $\int_0^\theta f(t)\,dt = 0$. Consider the solutions of the initial value problem

$$\varphi(u')' + f(u) = 0, \quad u(0) = \alpha, \quad u'(0) = 0.$$

If $\alpha > \theta$, then the conservation of energy formula (3.264) implies that

$$\frac{p-1}{p}|u'(x)|^p + F(u(x)) = F(\alpha) > 0,$$

and so this solution must vanish (for the first time) at some $r > 0$. Rescaling, we get a positive solution of (3.254) at $\lambda = r^2$.

We now use the Implicit Function Theorem to continue solutions for both decreasing and increasing λ (working, as before, in the space X from [11], or in the framework of B.P. Rynne [160]), and in both directions there are no turns by Lemma 3.41. For decreasing λ, we have $u(0, \lambda) \to \infty$, since otherwise the solution curve would have no place to go (it cannot go to zero, because $f(u)$ is negative near zero). By Lemma 2.46, we have an a priori bound for any $\lambda > 0$, and so $u(0, \lambda) \to \infty$ as $\lambda \to 0$. For increasing λ, $u(0, \lambda)$ is decreasing, in view of Lemma 3.42, and hence the oscillation of the solutions is decreasing. Consider first the case $f(0) = 0$. Then positive solutions continue for all λ, since $u_x(1, \lambda) < 0$, by Lemma 3.34. In case $f(0) < 0$, the opposite is true, i.e., we have $u_x(1, \lambda_0) = 0$ at some λ_0, and the solution becomes sign-changing for $\lambda > \lambda_0$, since otherwise (i.e., assuming that $u(x, \lambda) > 0$ for all λ) we shall get arbitrarily large oscillations of $u(x, \lambda)$ for large λ, by integrating the equation (3.254) over any interval $(x, 1)$. $\diamondsuit$

Bibliography

[1] Adamowicz T. and Korman P. (2011). Remarks on time map for quasilinear equations, *J. Math. Anal. Appl.* **376**, no. 2, pp. 686-695.

[2] Addou I. and Wang S.-H. (2003). Exact multiplicity results for a p-Laplacian problem with concave-convex-concave nonlinearities, *Nonlinear Anal.* **53**, no. 1, pp. 111-137.

[3] Adimurthi and Yadava S.L. (1994). An elementary proof of the uniqueness of positive radial solutions of a quasilinear Dirichlet problem, *Arch. Rational Mech. Anal.* **127**, no. 3, pp. 219-229.

[4] Adimurthy, Pacella F. and Yadava S.L. (1997). On the number of positive solutions for some semilinear Dirichlet problems in a ball, *Differential and Integral Equations* **10(6)**, pp. 1157-1170.

[5] Aduén H., Castro A., and Cossio J. (2008). Uniqueness of large radial solutions and existence of nonradial solutions for a superlinear Dirichlet problem in annulii, *J. Math. Anal. Appl.* **337**, no. 1, pp. 348-359.

[6] Aftalion A. and Pacella F. (2003). Uniqueness and nondegeneracy for some nonlinear elliptic problems in a ball, *J. Differential Equations* **195**, pp. 389-397.

[7] Allgower E. L. (1975). On a discretization of $y'' + \lambda y^k = 0$, *Topics in numerical analysis*, II (Proc. Roy. Irish Acad. Conf., Univ. College, Dublin, 1974), pp. 1-15. (Academic Press), London.

[8] Amann H. (1976). Fixed point equations and nonlinear eigenvalue problems in ordered Banach spaces, *SIAM Rev.* **18**, pp. 620-709.

[9] Ambrosetti A., Arcoya D. and Buffoni B. (1994). Positive solutions for some semi-positone problems via bifurcation theory, *Differential and Integral Equations* **7**, pp. 655-663.

[10] Ambrosetti A., Brezis H. and Cerami G. (1994). Combined effects of concave and convex nonlinearities in some elliptic problems, *J. Funct. Anal.* **122**, pp. 519-543.

[11] Ambrosetti A. and Prodi G. (1972). On the inversion of some differentiable mappings with singularities between Banach spaces, *Ann. Mat. Pura Appl.* **93 (4)**, pp. 231-246.

[12] Ambrosetti A. and Prodi G. (1993). A Primer of Nonlinear Analysis, (Cam-

bridge University Press).

[13] Anuradha V. and Shivaji R. (1992). Existence of infinitely many non-trivial bifurcation points, *Results in Math.* **22**, pp. 641-650.

[14] Arioli G., Gazzola F., Grunau H.-C. and Sassone E. (2008). The second bifurcation branch for radial solutions of the Brezis-Nirenberg problem in dimension four, *NoDEA Nonlinear Differential Equations Appl.* **15**, no. 1-2, pp. 69-90.

[15] Aris R. (1975). Mathematical Theory of Diffusion and Reaction in Permeable Catalysts, (Oxford University Press), London.

[16] Aubin T. (1976). Problmes isopérimétriques et espaces de Sobolev, (French) *J. Differential Geometry* **11**, no. 4, pp. 573-598.

[17] Bebernes J. and Eberly D. (1989). Mathematical Problems from Combustion Theory, (Springer-Verlag), New York.

[18] Bellman R. (1953). Stability Theory of Differential Equations, (McGraw-Hill).

[19] Berestycki H., Lions P.-L. and L. A. Peletier L. A. (1981). An ODE approach to the existence of positive solutions for semilinear problems in R^N, *Indiana Univ. Math. J.* **30**, no. 1, pp. 141-157.

[20] Berger M.S. and Podolak E. (1974/75). On the solutions of a nonlinear Dirichlet problem, *Indiana Univ. Math. J.* **24**, pp. 837-846.

[21] Binding P.A. and Rynne B.P. (2007). The spectrum of the periodic *p*-Laplacian, *J. Differential Equations* **235**, no. 1, pp. 199-218.

[22] Bonheure D., Habets P., Obersnel F., and Omari P. (2007). Classical and non-classical solutions of a prescribed curvature equation, *J. Differential Equations* **243** no. 2, pp. 208-237.

[23] Bratu G. (1914). Sur les equations integrales non lineares, *Bull. Soc. Math. de France* **42**, pp. 113-142.

[24] Brock F. (1998). Radial symmetry for nonnegative solutions of semilinear elliptic equations involving the *p*-Laplacian. Progress in partial differential equations, Vol. 1 (Pont--Mousson, 1997), pp. 46-57, (Pitman Res. Notes Math. Ser.), **383**, Longman.

[25] Brown K.J. (2000). Curves of solutions through points of neutral stability, *Result. Math.* **37**, pp. 315-321.

[26] Brown K.J., Ibrahim M. M. A. and Shivaji R. (1981). *S*-shaped bifurcation curves, *Nonlinear Anal.* **5**, no. 5, pp. 475-486.

[27] Budd C. and Norbury J. (1987). Semilinear elliptic equations and supercritical growth, *J. Differential Equations* **68** no. 2, pp. 169-197.

[28] Cañada A. (2007). Multiplicity results near the principal eigenvalue for boundary-value problems with periodic nonlinearity, *Math. Nachr.* **280**, no. 3, pp. 235-241.

[29] Castro A., Gadam S. and Shivaji R. (1997). Positive solution curves of semipositone problems with concave nonlinearities, *Proc. Roy. Soc. Edinburgh Sect. A* **127**, no. 5, pp. 921-934.

[30] Castro A. and Shivaji R. (1989). Nonnegative solutions to a semilinear Dirichlet problem in a ball are positive and radially symmetric, *Comm. Partial Differential Equations* **14**, no. 8-9, pp. 1091-1100.

[31] Catrina F. (2009). A note on a result of M. Grossi, *Proc. Amer. Math. Soc.* **137**, no. 11, pp. 3717-3724.

[32] Cerami G. (1986). Symmetry breaking for a class of semilinear elliptic problems, *Nonlinear Analysis TMA* **10(1)**, pp. 1-14.

[33] Chang K.C. (1993). Infinite-dimensional Morse theory and multiple solution problems, *Progress in Nonlinear Differential Equations and their Applications, 6.* (Birkhauser) Boston.

[34] Chen T.F., Levine H.A. and Sacks P.E. (1988). Analysis of a convective reaction-diffusion equation, *Nonlinear Analysis TMA* **12**, pp. 1349-1370.

[35] Cheng J. (2005). Uniqueness results for the one-dimensional p-Laplacian, *J. Math. Anal. Appl.* **311**, no. 2, pp. 381-388.

[36] Clément P., de Figueiredo D. G. and Mitidieri E. (1992). Positive solutions of semilinear elliptic systems, *Comm. Partial Differential Equations* **17** , no. 5-6, pp. 923-940.

[37] Costa D., Jeggle H., Schaaf R. and Schmitt K. (1988). Oscillatory perturbations of linear problems at resonance, *Results in Mathematics* **14**, pp. 275-287.

[38] Courant R. and Hilbert D. (1953). Methods of Mathematical Physics, vol. 1, (Interscience), New York.

[39] Crandall M.G. and Rabinowitz P.H. (1971). Bifurcation from simple eigenvalues, *Journal of Functional Analysis*, **8**, pp. 321-340.

[40] Crandall M.G. and Rabinowitz P.H. (1973). Bifurcation, perturbation of simple eigenvalues and linearized stability, *Arch. Rational Mech. Anal.*, **52**, pp. 161-180.

[41] Dalmasso R. (1993). Symmetry properties of solutions of some fourth order ordinary differential equations, *Bull. Sci. Math.* **117**, pp. 441-462.

[42] Dalmasso R. (1996). Uniqueness of positive solutions for some nonlinear fourth-order equations, *J. Math. Anal. Appl.* **201**, pp. 152-168.

[43] Dalmasso R. (2000). Existence and uniqueness of positive solutions of semilinear elliptic systems, *Nonlinear Anal.* **39**, no. 5, Ser. A: Theory Methods, pp. 559-568.

[44] Damascelli L., Grossi M. and Pacella F. (1999). Qualitative properties of positive solutions of semilinear elliptic equations in symmetric domains via the maximum principle, *Ann. Inst. H. Poincare Anal. Non Lineaire* **16**, no. 5, pp. 631-652.

[45] Damascelli L. and Pacella F. (1998). Monotonicity and symmetry of solutions of p-Laplace equations, $1 < p < 2$, via the moving plane method, *Ann. Scuola Norm. Sup. Pisa Cl. Sci.* **26**, no. 4, pp. 689-707.

[46] Dancer E.N. (1979). On non-radially symmetric bifurcation, *J. London Math. Soc.* **20**, pp. 287-292.

[47] Dancer E.N. (1980). On the structure of solutions of an equation in catalysis theory when a parameter is large, *J. Differential Equations* **37**, no. 3, pp. 404-437.

[48] Dancer E.N. (1990). The effect of domain shape on the number of positive solutions of certain nonlinear equations II, *J. Differential Equations* **87**, pp. 316-339.

[49] Ding Z., Chen G. and Li S. (2005). On positive solutions of the elliptic sine-Gordon equation, *Commun. Pure Appl. Anal.* **4**, no. 2, pp. 283-294.

[50] Drabek P. (2007). The p-Laplacian - mascot of nonlinear analysis, *Acta Math. Univ. Comenian. (N.S.)* **76**, no. 1, pp. 85-98.

[51] Du Y. and Lou Y. (2001). Proof of a conjecture for the perturbed Gelfand equation from combustion theory, *J. Differential Equations* **173**, pp. 213-230.

[52] Dunninger D.R. (2001). Multiplicity of positive solutions for a nonlinear fourth order equation, *Ann. Polon. Math.* **77**, pp. 161-168.

[53] Erbe L. and Tang M. M. Structure of positive radial solutions of semilinear elliptic equations, *J. Differential Equations* **133**, no. 2, pp. 179-202.

[54] Evans L. (1998). Partial Differential Equations. Graduate Studies in Mathematics, 19. (American Mathematical Society), Providence, RI.

[55] de Figueiredo D.G. (2008). Semilinear elliptic systems: existence, multiplicity, symmetry of solutions, Handbook of Differential Equations, Stationary Partial Differential Equations, Vol. **5**, Edited by M. Chipot,(Elsevier Science, North Holland), pp. 1-48.

[56] de Figueiredo D.G. and Ni W.-M. (1979). Perturbations of second order linear elliptic problems by nonlinearities without Landesman-Lazer condition, *Nonlinear Anal.* **3** no. 5, pp. 629-634.

[57] Fleckinger J. and Reichel W. (2005). Global solution branches for p-Laplacian boundary value problems, *Nonlinear Anal.* **62**, no. 1, pp. 53-70.

[58] Flores I. (2004). A resonance phenomenon for ground states of an elliptic equation of Emden-Fowler type, *J. Differential Equations* **198**, no. 1, pp. 1-15.

[59] Franchi B., Lanconelli E. and Serrin J. (1996). Existence and uniqueness of nonnegative solutions of quasilinear equations in R^n, *Adv. Math.* **118**, no. 2, pp. 177-243.

[60] Galstian A., Korman P. and Li Y. (2006). On the oscillations of the solution curve for a class of semilinear equations, *J. Math. Anal. Appl.* **321**, no. 2, pp. 576-588.

[61] Gardner R. and Peletier L.A. (1986). The set of positive solutions of semilinear equations in large balls, *Proc. Roy. Soc. Edinburgh Sect. A* **104**, no. 1-2, pp. 53-72.

[62] Gaudenzi M., Habets P. and Zanolin F. (2004). A seven-positive-solutions theorem for a superlinear problem, *Adv. Nonlinear Stud.* **4**, no. 2, pp. 149-164.

[63] Gidas B., Ni W.-M. and Nirenberg L. (1979). Symmetry and related properties via the maximum principle, *Commun. Math. Phys.* **68**, pp. 209-243.

[64] Gidas B. and Spruck J. (1981). A priori bounds for positive solutions of nonlinear elliptic equations, *Comm. Partial Differential Equations* **6**, no. 8, pp. 883-901.

[65] Grossi M. (2008). Radial solutions for the Brezis-Nirenberg problem involving large nonlinearities, *J. Funct. Anal.* **254**, no. 12, pp. 2995-3036.

[66] Guo Z. and Wei J. (2008). Infinitely many turning points for an elliptic problem with a singular non-linearity, *J. Lond. Math. Soc.* **78**, no. 1, pp.

21-35.

[67] Habets P. and Omari P. (2007). Multiple positive solutions of a one-dimensional prescribed mean curvature problem, *Commun. Contemp. Math.* **9**, no. 5, pp. 701-730.

[68] Hastings S.P. and J. B. McLeod J. B. (1985). The number of solutions to an equation from catalysis, *Proc. Roy. Soc. Edinburgh Sect. A* **101**, no. 1-2, pp. 15-30.

[69] J. Hempel J. (1977). On a superlinear differential equation, *Indiana Univ. Math. J.* **26**, pp. 265-275.

[70] Henry D. (1981). Geometric Theory of Semilinear Parabolic Equations, Lecture Notes in Math., Vol. **840**, (Springer), New York.

[71] Hochstadt H. (1986). The functions of mathematical physics. Second edition. With a foreword by Wilhelm Magnus. (Dover Publications, Inc.), New York.

[72] Holzmann M. and Kielhofer H. (1994). Uniqueness of global positive solution branches of nonlinear elliptic problems, *Math. Ann.* **300**, no. 2, pp. 221-241.

[73] Hung K.-C. and Wang S.-H. A theorem on S-shaped bifurcation curve for a positone problem with convex-concave nonlinearity and its applications to the perturbed Gelfand problem, Preprint.

[74] Iaia J.A. (2006). Localized solutions of elliptic equations: loitering at the hilltop, *Electron. J. Qual. Theory Differ. Equ.*, No. 12, 15 pp. (electronic)

[75] Joseph D.D. and Lundgren T.S. (1972/73). Quasilinear Dirichlet problems driven by positive sources, *Arch. Rational Mech. Anal.* **49**, pp. 241-269.

[76] (1993). Kielhofer H. and Maier S. Infinitely many positive solutions of semilinear elliptic problems via sub- and supersolutions, *Comm. Partial Differential Equations* **18**, pp. 1219-1229.

[77] Korman P. (1989). A maximum principle for fourth order ordinary differential equations, *Appl. Anal.* **33**, no. 3-4, pp. 267-273.

[78] Korman P. (1995). Symmetry of positive solutions for elliptic problems in one dimension, *Appl. Anal.* **58**, no. 3-4, pp. 351-365.

[79] Korman P. (1997). Solution curves for semilinear equations on a ball, *Proc. Amer. Math. Soc.* **125**, pp. 1997-2005.

[80] Korman P. (1997). Steady states and long time behavior of some convective reaction-diffusion equations, *Funkcialaj Ekvacioj* **40**, pp. 165-183.

[81] Korman P. (1998). The global solution set for a class of semilinear problems, *J. Math. Anal. Appl.* **226**, no. 1, pp. 101-120.

[82] Korman P. (1999). Multiplicity of positive solutions for semilinear equations on circular domains, *Comm. Appl. Nonlinear Anal.* **6**, no. 3, pp. 17-35.

[83] Korman P. (2001). On the multiplicity of solutions of semilinear equations, *Math. Nachr.* **229**, pp. 119-127.

[84] Korman P. (2002). Curves of sign-changing solutions for semilinear equations, *Nonlinear Anal. TMA* **51**, no. 5, pp. 801-820.

[85] Korman P. (2002). On uniqueness of positive solutions for a class of semilinear equations, *Discrete Contin. Dyn. Syst.* **8**, no. 4, pp. 865-871.

[86] Korman P. (2002). A remark on the non-degeneracy condition, *Result.*

Math. **41**, pp. 334-336.

[87] Korman P. (2002). Global solution curves for semilinear systems, *Math. Methods Appl. Sci.* **25**, no. 1, pp. 3-20.

[88] Korman P. (2003). Curves of positive solutions for supercritical problems, *Appl. Anal.* **82**, no. 1, pp. 45-54.

[89] Korman P. (2003). An accurate computation of the global solution curve for the Gelfand problem through a two point approximation, *Appl. Math. Comput.* **139**, no. 2-3, pp. 363-369.

[90] Korman P. (2004). Further remarks on the non-degeneracy condition, *Result. Math.* **45**, pp. 293-298.

[91] Korman P. (2004). Uniqueness and exact multiplicity of solutions for a class of fourth-order semilinear problems, *Proc. Roy. Soc. Edinburgh Sect. A* **134**, no. 1, pp. 179-190.

[92] Korman P. (2006). Global solution branches and exact multiplicity of solutions for two point boundary value problems, Handbook of Differential Equations, Ordinary Differential Equations, Vol. **3**, Edited by A. Canada, P. Drabek and A. Fonda, (Elsevier Science, North Holland), pp. 547-606.

[93] Korman P. (2006). Uniqueness and exact multiplicity of solutions for non-autonomous Dirichlet problems, *Adv. Nonlinear Stud.* **6**, no. 3, pp. 461-481.

[94] Korman P. (2007). Stability and instability of solutions of semilinear problems, *Appl. Anal.* **86**, no. 2, pp. 135-147.

[95] Korman P. (2007). Computation of radial solutions of semilinear equations, *Electron. J. Qual. Theory Differ. Equ.*, No. 13, 14 pp. (electronic).

[96] Korman P. (2008). Uniqueness and exact multiplicity of solutions for a class of Dirichlet problems, *J. Differential Equations* **244**, no. 10, pp. 2602-2613.

[97] Korman P. (2008). An oscillatory bifurcation from infinity, and from zero, *NoDEA Nonlinear Differential Equations Appl.* **15**, no. 3, pp. 335-345.

[98] Korman P. (2008). A global approach to ground state solutions, *Electron. J. Differential Equations*, No. 122, 13 pp.

[99] Korman P. (2009). Curves of equiharmonic solutions, and ranges of nonlinear equations, *Adv. Differential Equations* **14**, no. 9-10, pp. 963-984.

[100] Korman P. (2010). Global solution curves for a class of elliptic systems, *Applicable Analysis* **89**, no. 6, pp. 849-860.

[101] Korman P. (2011). Existence and uniqueness of solutions for a class of *p*-Laplace equations on a ball, *Adv. Nonlinear Stud.* **11**, pp. 875-888.

[102] Korman P. Nonlinear perturbations of linear elliptic systems at resonance, *Proc. Amer. Math. Soc.* In press.

[103] Korman P. and Li Y. (1999). On the exactness of an *S*-shaped bifurcation curve, *Proc. Amer. Math. Soc.* **127**, no. 4, pp. 1011-1020.

[104] Korman P. and Li Y. (2000). Infinitely many solutions at a resonance, *Electron. J. Diff. Eqns.*, **Conf. 05**, pp. 105-111.

[105] Korman P. and Li Y. (2010). Computing the location and the direction of bifurcation for sign changing solutions, *Differential Equations & Applications* **2**, no. 1, pp. 1-13.

[106] Korman P. and Li Y. (2010). Global solution curves for a class of quasilinear boundary value problems, *Proc. Roy. Soc. Edinburgh Sect. A* **140**, no. 6,

pp. 1197-1215.

[107] Korman P. and Li Y. A computer assisted study of uniqueness of ground state solution, Preprint.

[108] Korman P., Li Y. and Ouyang T. (1996). Exact multiplicity results for boundary-value problems with nonlinearities generalising cubic, *Proc. Royal Soc. Edinburgh, Ser. A* **126A**, pp. 599-616.

[109] Korman P., Li Y. and Ouyang T. (1997). An exact multiplicity result for a class of semilinear equations, *Comm. Partial Differential Equations* **22**, no. 3-4, pp. 661-684.

[110] Korman P., Li Y. and Ouyang T. (2003). Perturbation of global solution curves for semilinear problems, *Adv. Nonlinear Stud.* **3**, no. 2, pp. 289-299.

[111] Korman P., Li Y. and Ouyang T. (2005). Computing the location and direction of bifurcation, *Math. Res. Lett.* **12**, no. 5-6, pp. 933-944.

[112] Korman P. and Ouyang T. (1996). Solution curves for two classes of boundary-value problems, *Nonlinear Anal.* **27**, no. 9, pp. 1031-1047.

[113] Korman P. and Ouyang T. (1993). Exact multiplicity results for two classes of boundary value problems, *Differential Integral Equations* **6**, no. 6, pp. 1507-1517.

[114] Korman P. and Ouyang T. (1995). Multiplicity results for two classes of boundary-value problems, *SIAM J. Math. Anal.* **26**, no. 1, pp. 180-189.

[115] Korman P. and Ouyang T. (1995). Exact multiplicity results for a class of boundary-value problems with cubic nonlinearities, *J. Math. Anal. Appl.* **194**, no. 1, pp. 328-341.

[116] Korman P. and Ouyang T. (2006). Positivity for the linearized problem for semilinear equations, *Topol. Methods Nonlinear Anal.* **28**, no. 1, pp. 53-60.

[117] Korman P. and Shi J. (2000). Instability and exact multiplicity of solutions of semilinear equations. Proceedings of the Conference on Nonlinear Differential Equations (Coral Gables, FL, 1999), 311-322 (electronic), Electron. J. Differ. Equ. Conf., 5, Southwest Texas State Univ., San Marcos, TX.

[118] Korman P. and Shi J. (2001). New exact multiplicity results with an application to a population model, *Proc. Roy. Soc. Edinburgh Sect. A* **131**, no. 5, pp. 1167-1182.

[119] Korman P. and Shi J. (2005). On Lane-Emden type systems, *Discrete Contin. Dyn. Syst.*, suppl., pp. 510-517.

[120] Kwong M.K. (1989). Uniqueness of positive solutions of $\Delta u - u + u^p = 0$ in R^n, *Arch. Rational Mech. Anal.* **105**, no. 3, pp. 243-266.

[121] Kwong M.K. and Li Y. (1992). Uniqueness of radial solutions of semilinear elliptic equations, *Trans. Amer. Math. Soc.* **333**, no. 1, pp. 339-363.

[122] Kwong M.K. and Zhang L. (1991). Uniqueness of the positive solution of $\Delta u + f(u) = 0$ in an annulus, *Differential Integral Equations* **4** no. 3, pp. 583-599.

[123] Landesman E.M. and Lazer A.C. (1970). Nonlinear perturbations of linear elliptic boundary value problems at resonance, *J. Math. Mech.* **19**, pp. 609-623.

[124] Leighton W. and Nehari Z. (1958). On the oscillation of solutions of self-adjoint linear differential equations of the fourth order, *Trans. Amer. Math.*

Soc. **89**, pp. 325-377.

[125] Levine H.A., Payne L.E., Sacks P.E. and Straughan B. (1989). Analysis of a convective reaction-diffusion equation II, *SIAM J. Math. Anal.* **20** pp. 133-147.

[126] Lin C.S. (1994). Uniqueness of least energy solutions to a semilinear elliptic equation in R^2, *Manuscripta Mathematica* **84**, pp. 13-19.

[127] Lin C.S. and Ni W.-M. (1988). A counterexample to the nodal domain conjecture and related semilinear equation, *Proc. Amer. Math. Soc.* **102**, pp. 271-277.

[128] Lin S.S. (1990). Positive radial solutions and non-radial bifurcation for semilinear elliptic equations in annular domains, *J. Differential Equations* **86**, pp. 367-391.

[129] Lindqvist P. (2006). Notes on the p-Laplace equation. Report. University of Jyväskylä Department of Mathematics and Statistics, 102. University of Jyväskylä, Jyväskylä.

[130] Lions P.-L. (1982). On the existence of positive solutions of semilinear elliptic equations, *SIAM Rev.* **24**, no. 4, pp. 441-467.

[131] McKean H.P. and Scovel J.C. (1986). Geometry of some simple nonlinear differential operators, *Ann. Scuola Norm. Sup. Pisa Ser. iv* **13**, pp. 299-346.

[132] McKenna P. J., Pacella F., Plum M. and Roth D. (2009). A uniqueness result for a semilinear elliptic problem: a computer assisted proof, *J. Differential Equations* **247**, pp. 2140-2162.

[133] McLeod K. and Serrin J. (1987). Uniqueness of positive radial solutions of $\Delta u + f(u) = 0$ in R^n, *Arch. Rational Mech. Anal.* **99**, pp. 115-145.

[134] Merle F. and Peletier L.A. (1991). Positive solutions of elliptic equations involving supercritical growth, *Proc. Roy. Soc. Edinburgh Sect. A* **118**, no. 1-2, pp. 49-62.

[135] Mitidieri E. (1993). A Rellich type identity and applications, *Comm. Partial Differential Equations* **18**, no. 1-2, pp. 125-151.

[136] Nagasaki K. and Suzuki T. (1994). Spectral and related properties about the Emden-Fowler equation $-\Delta u = \lambda e^u$ on circular domains, *Math. Ann.* **299**, no. 1, pp. 1-15.

[137] Ni W.-M. (1983). Uniqueness of solutions of nonlinear Dirichlet problems, *J. Differential Equations* **50**, no. 2, pp. 289-304.

[138] Ni W.-M. and Nussbaum R.D. (1985). Uniqueness and nonuniqueness of positive radial solutions of $\Delta u + f(u, r)$, *Commun. Pure Appl. Math.* **38**, pp. 69-108.

[139] Nirenberg L. (1974). Topics in Nonlinear Functional Analysis, Courant Institute Lecture Notes (Amer. Math. Soc.).

[140] Nirenberg L. Letter to P. Korman of 03/13/2003.

[141] Opial Z. (1961). Sur les periodes des solutions de l'equation differentielle $x'' + g(x) = 0$, (French) *Ann. Polon. Math.* **10**, pp. 49-72.

[142] Ouyang T. (1992). On the positive solutions of semilinear equations $\Delta u + \lambda u - h u^p = 0$ on the compact manifolds, *Trans. Amer. Math. Soc.* **331**, no. 2, pp. 503-527.

[143] Ouyang T. (1991). On the positive solutions of semilinear equations $\Delta u +$

$\lambda u + hu^p = 0$ on compact manifolds, II, *Indiana Univ. Math. J.* **40**, no. 3, pp. 1083-1141.

[144] Ouyang T. and J. Shi, Exact multiplicity of positive solutions for a class of semilinear problems. *J. Differential Equations* **146**, pp. 121-156 (1998).

[145] Ouyang T. and J. Shi, Exact multiplicity of positive solutions for a class of semilinear problems, II, *J. Differential Equations* **158**, no. 1, pp. 94-151 (1999).

[146] Pacard F. and Rivière T. (2000). Linear and Nonlinear Aspects of Vortices. The Ginzburg-Landau model. Progress in Nonlinear Differential Equations and their Applications, 39. (Birkhäuser), Boston.

[147] Pan H. (2009). One-dimensional prescribed mean curvature equation with exponential nonlinearity, *Nonlinear Anal.* **70** pp. 999-1010.

[148] Peletier L.A. and Serrin J. (1983). Uniqueness of positive solutions of semilinear equations in R^n, *Arch. Rational Mech. Anal.* **81** no. 2, pp. 181-197.

[149] Peletier L.A. and Serrin J. (1986). Uniqueness of nonnegative solutions of semilinear equations in R^n, *J. Differential Equations* **61** no. 3, pp. 380-397.

[150] Peletier L.A. and Van der Vorst R. C. A. M. (1992). Existence and nonexistence of positive solutions of nonlinear elliptic systems and the biharmonic equation, *Differential Integral Equations* **5**, no. 4, pp. 747-767.

[151] Pohozaev S.I. (1965). On the eigenfunctions of the equation $\Delta u + \lambda f(u) = 0$. (Russian) *Dokl. Akad. Nauk SSSR* **165**, pp. 36-39.

[152] Pucci P. and Serrin J. (1998). Uniqueness of ground states for quasilinear elliptic operators, *Indiana Univ. Math. J.* **47** no. 2, pp. 501-528.

[153] Pucci P. and Serrin J. (1986). A general variational identity, *Indiana Univ. Math. J.* **35**, no. 3, pp. 681-703.

[154] Quittner P. and Souplet P. (2007). Superlinear Parabolic Problems. Blowup, global existence and steady states. Birkhäuser Advanced Texts: Basler Lehrbücher. (Birkhäuser), Basel.

[155] Ramaswamy M. (1986). On the global set of solutions of a nonlinear ODE: theoretical and numerical description, *J. Differential Equations*, **65**, pp. 1-48.

[156] Ramaswamy M. and Srikanth P.N. (1987). Symmetry breaking for a class of semilinear elliptic problems, *Trans. Amer. Math. Soc.* **304**, pp. 839-845.

[157] Renardy M. and Rogers R.C. (1993). An Introduction to Partial Differential Equations. Texts in Applied Mathematics, 13. (Springer-Verlag), New York.

[158] Ruf B. (2008). Superlinear elliptic equations and systems, Handbook of Differential Equations, Stationary Partial Differential Equations, Vol. **5**, Edited by M. Chipot, (Elsevier Science), North Holland, pp. 277-370.

[159] Rynne B.P. (2004). Solution curves of $2m$-th order boundary-value problems, *Electron. J. Differential Equations*, No. 32, 16 pp. (electronic).

[160] Rynne B.P. (2010). A global curve of stable, positive solutions for a p-Laplacian problem, *Electron. J. Differential Equations*, No. 58, 12 pp.

[161] Schaaf R. (1990). Global Solution Branches of Two Point Boundary Value Problems, Lecture Notes in Mathematics, no. **1458**, (Springer-Verlag).

[162] Schaaf R. and Schmitt K. (1988). A class on nonlinear Sturm-Liouville problems with infinitely many solutions, *Trans. Amer. Math. Soc.* **306**, no.

2, pp. 853-859.

[163] Schaaf R. and Schmitt K. (1992). Asymptotic behavior of positive solution branches of elliptic problems with linear part at resonance, *Z. Angew. Math. Phys.* **43** no. 4, pp. 645-676.

[164] Schmitt K. (1995). Positive solutions of semilinear elliptic boundary value problems. Topological methods in differential equations and inclusions (Montreal, PQ, 1994), pp. 447-500, NATO Adv. Sci. Inst. Ser. C Math. Phys. Sci., 472, (Kluwer Acad. Publ.), Dordrecht.

[165] Schmitt K., Thompson R.C. and Walter W. (1978). Existence of solutions of a nonlinear boundary value problem via the method of lines, *Nonlinear Anal.* **2**, no. 5, pp. 519-535.

[166] Serrin J. A symmetry problem in potential theory, *Arch. Rational Mech. Anal.* **43**, pp. 304-318 (1971).

[167] Serrin J. and Tang M. (2000). Uniqueness of ground states for quasilinear elliptic equations, *Indiana Univ. Math. J.* **49**, no. 3, pp. 897-923.

[168] Serrin J. and H. Zou, (2002). Cauchy-Liouville and universal boundedness theorems for quasilinear elliptic equations and inequalities, *Acta Math.* **189**, no. 1, pp. 79-142.

[169] Shi J. (1999). Persistence and bifurcation of degenerate solutions, *J. Funct. Anal.* **169**, no. 2, pp. 494-531.

[170] Shi J. (2002). Semilinear Neumann boundary value problems on a rectangle, *Trans. Amer. Math. Soc.* **354**, no. 8, pp. 3117-3154.

[171] Shi J. and Wang J. (1999). Morse indices and exact multiplicity of solutions to semilinear elliptic problems, *Proc. Amer. Math. Soc.* **127**, no. 12, pp. 3685-3695.

[172] Sidorov Y.V., Fedoryuk M.V. and Shabunin M.I. (1985). *Lectures on the Theory of Functions of Complex Variable,* (Mir Publishers), Moscow.

[173] Smets D., Willem M. and Su J. (2002). Non-radial ground states for the Hénon equation, *Commun. Contemp. Math.* **4**, no. 3, pp. 467-480.

[174] Smoller J. and Wasserman A. (1981). Global bifurcation of steady-state solutions, *J. Differential Equations* **39**, pp. 269-290.

[175] Smoller J. and Wasserman A. (1986). Symmetry-breaking for positive solutions of semilinear elliptic problems, *Arch. Rat. Mech. Anal.* **95**, pp. 217-225.

[176] Smoller J. and Wasserman A. (1988). Bifurcation from symmetry, In *Nonlinear Diffusion Equations and Their Equilibrium States,* W.-M. Ni, L.A. Peletier and J. Serrin, Editors, (Springer-Verlag).

[177] Sperb R. (1981). Maximum Principles and Their Applications. Mathematics in Science and Engineering, **157**. (Academic Press, Inc.), New York-London.

[178] Srikanth P.N. (1993). Uniqueness of solutions of nonlinear Dirichlet problems, *Differential Integral Equations* **6**, no. 3, pp. 663-670.

[179] Talenti G. (1976). Best constant in Sobolev inequality, *Ann. Mat. Pura Appl.* **110 (4)**, pp. 353-372.

[180] Tanaka S. (2008). Uniqueness of nodal radial solutions of superlinear elliptic equations in a ball, *Proc. Roy. Soc. Edinburgh Sect. A* **138**, no. 6, pp. 1331-

1343.

[181] Tang M. (2000). Uniqueness of positive radial solutions for n-Laplacian Dirichlet problems, *Proc. Roy. Soc. Edinburgh Sect. A* **130**, no. 6, pp. 1405-1416.

[182] Tang M. (2003). Uniqueness of positive radial solutions for $\Delta u - u + u^p = 0$ on an annulus, *J. Differential Equations* **189**, no. 1, pp. 148-160.

[183] Tang M. (2003). Exact multiplicity for semilinear elliptic Dirichlet problems involving concave and convex nonlinearities, *Proc. Roy. Soc. Edinburgh Sect. A* **133**, no. 3, pp. 705-717.

[184] Troy W.C. (1981). Symmetry properties in systems of semilinear elliptic equations, *J. Differential Equations* **42**, no. 3, pp. 400-413.

[185] Vanderbauwhede A. (1982). Local Bifurcation and Symmetry, Research Notes in Mathematics, No. 75, (Pitman).

[186] Van der Vorst R.C.A.M. (1992). Variational identities and applications to differential systems, *Arch. Rational Mech. Anal.* **116**, no. 4, pp. 375-398.

[187] Vázquez J.L. (1984). A strong maximum principle for some quasilinear elliptic equations, *Appl. Math. Optim.* **12**, no. 3, pp. 191-202.

[188] Wang S.-H. (1989). A correction for a paper by J. Smoller and A. Wasserman, *J. Differential Equations* **77**, pp. 199-202.

[189] Wang S.-H. (1994). On S-shaped bifurcation curves, *Nonlinear Analysis, TMA,* **22**, pp. 1475-1485.

[190] Wang S.-H. (1998). Rigorous analysis and estimates of S-shaped bifurcation curves in a combustion problem with general Arrhenius reaction-rate laws, *R. Soc. Lond. Proc. Ser. A Math. Phys. Eng. Sci.* **454**, no. 1972, pp. 1031-1048.

[191] Wang S.-H. and Lee F.P. (1994). Bifurcation of an equation from catalysis theory, *Nonlinear Anal.* **23**, no. 9, pp. 1167-1187.

[192] Wang S.-H. and Yeh T.-S. (2008). Exact multiplicity of solutions and S-shaped bifurcation curves for the p-Laplacian perturbed Gelfand problem in one space variable, *J. Math. Anal. Appl.* **342**, no. 2, pp. 1175-1191.

[193] Yanagida E. (1991). Uniqueness of positive radial solutions of $\Delta u + g(r)u + h(r)u^p = 0$ in R^n, *Arch. Rational Mech. Anal.* **115**, no. 3, pp. 257-274.

[194] Zeidler E. (1986). *Nonlinear Functional Analysis and its Applications*, vol. I, (Springer).

[195] Zhang L. (1992). Uniqueness of positive solutions of $\Delta u + u + u^p = 0$ in a ball, *Comm. Partial Differential Equations* **17**, no. 7-8, pp. 1141-1164.

[196] Zhao P. and Zhong C. (2002). Positive solutions of elliptic equations involving both supercritical and sublinear growth, *Houston J. Math.* **28**, no. 3, pp. 649-663.